## In addition to school hours, how many hours do you study per day from Mondays to Fridays?

(including cram schools and time with a tutor)

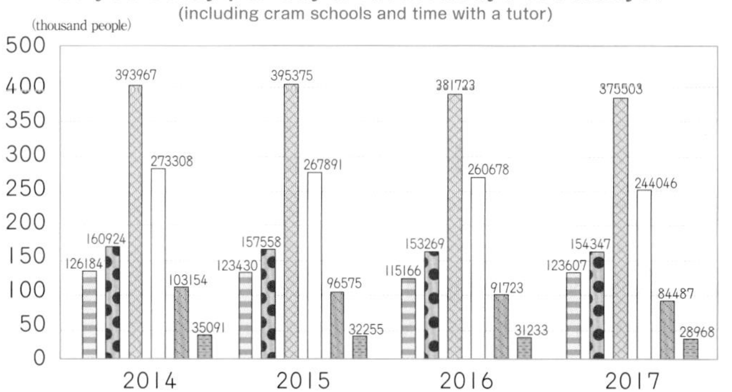


## How many hou Saturdays, Su

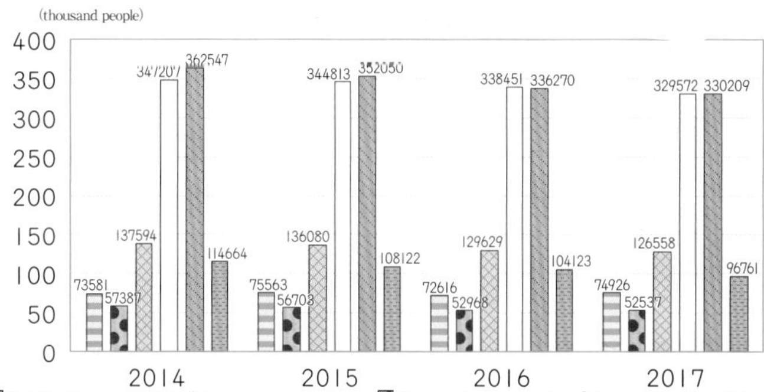


## Do you like studying Japanese?

(thousand people)

| | It applies | Somewhat applies | Somewhat doesn't apply | Doesn't apply |
|---|---|---|---|---|
| 2014 | 252761 | 397219 | 288784 | 153577 |
| 2015 | 268368 | 390011 | 269860 | 143921 |
| 2016 | 247594 | 357780 | 269595 | 156370 |
| 2017 | 259832 | 353529 | 255929 | 140955 |

## Do you think that studying Japanese is important?

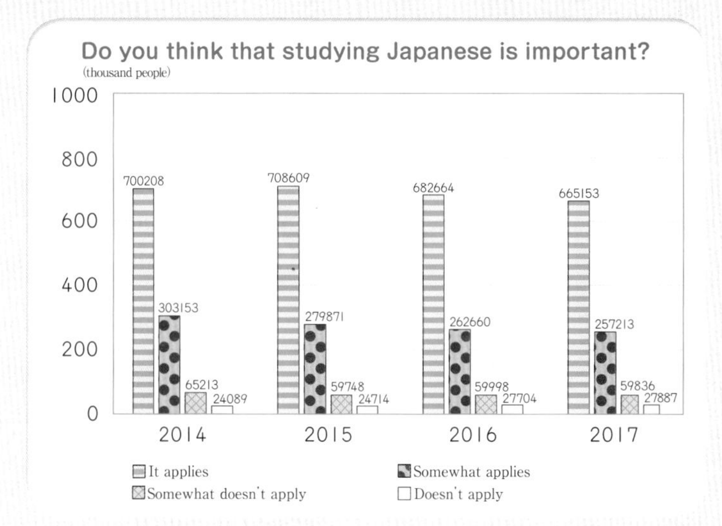


## Do you like studying mathematics?

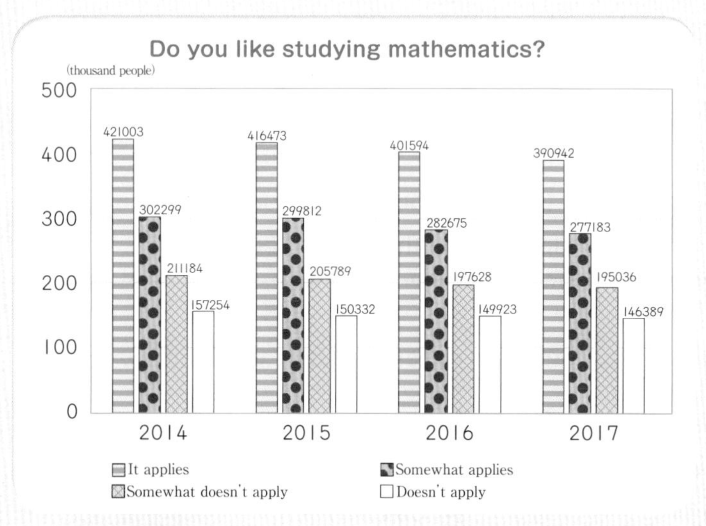


## Do you think that studying mathematics is important?

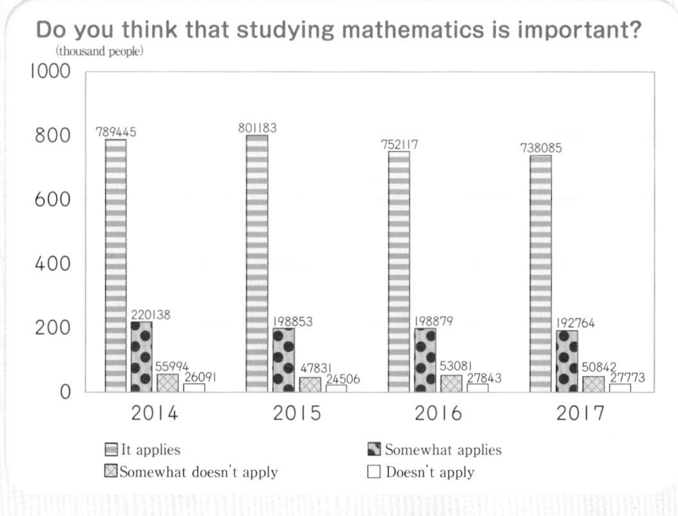

Yui: It's decreasing overall.

Daiki: Since the number of children is different, is it possible to say that it is decreasing?

01502

# Table of contents

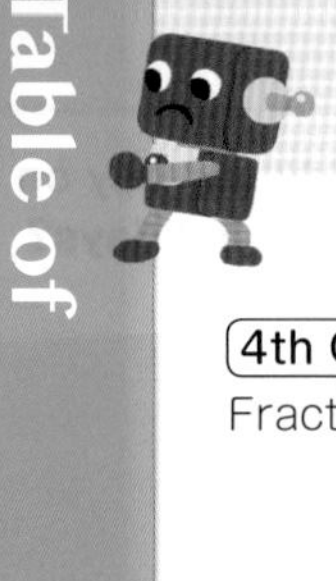



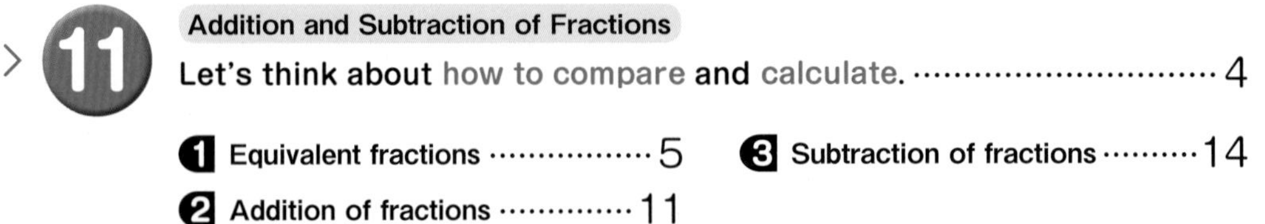
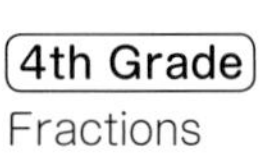

Hiroto

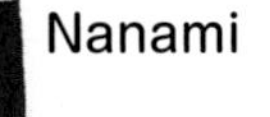

Nanami

Daiki

Let's learn mathematics together.

Yui

Important words and rules

Let's deepen.

You will want to learn much more.

Rules that you found

Want to connect

Solve new problems.

Find the ?

# Which one is larger?

Let's play the fraction card game.

## How to play

Pick up one card that has a fraction written in it. The one with the largest fraction card wins.

1

I win because mine is larger.
5/6
3/6
2

Since the numerator is the same, the fraction with the smaller denominator is larger.

$\frac{2}{3}$ $\frac{2}{4}$

3

With these cards, who wins?
?
2/3
6/9
4

**Problem** How can we compare the size of fractions that have different denominators?

Addition and Subtraction of Fractions

# Let's think about how to compare and calculate.

## 1 Equivalent fractions

Want to explore

**1** **Let's look at the number line below and explore fractions that are equivalent to $\frac{1}{2}$.**

① Let's write the number that applies in the following □.

$$\frac{1}{2} = \frac{\square}{4} = \frac{\square}{6} = \frac{\square}{8} = \frac{5}{\square} = \frac{6}{\square}$$

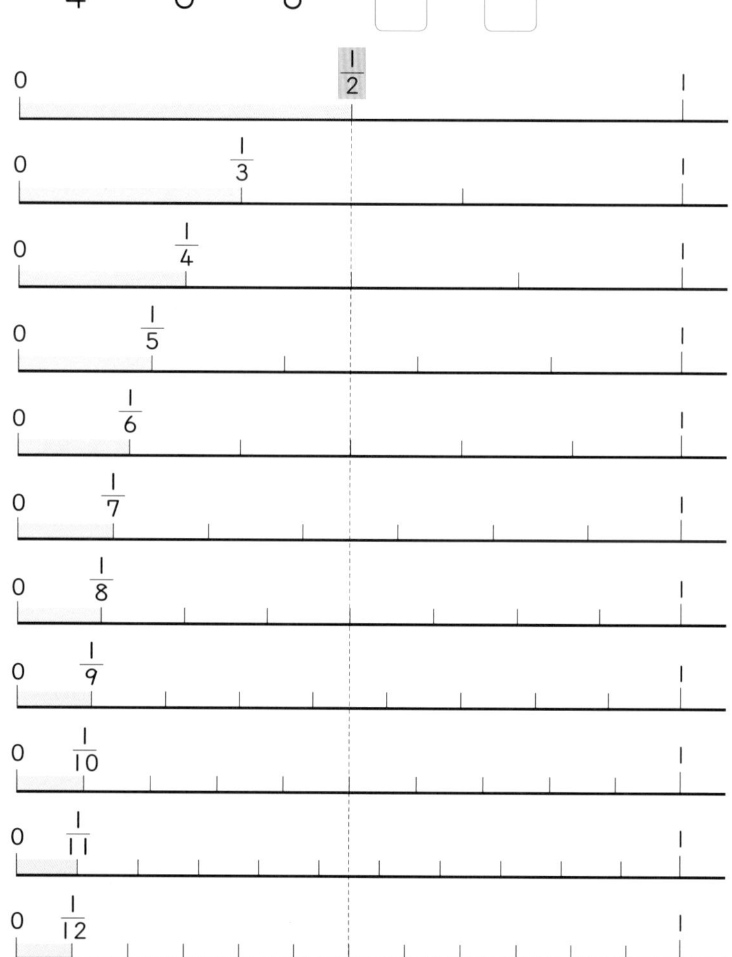


Purpose As for equivalent fractions, what kind of rules are there?

② Each of the following numerators and denominators is how many times of $\frac{1}{2}$ in exercise ①?

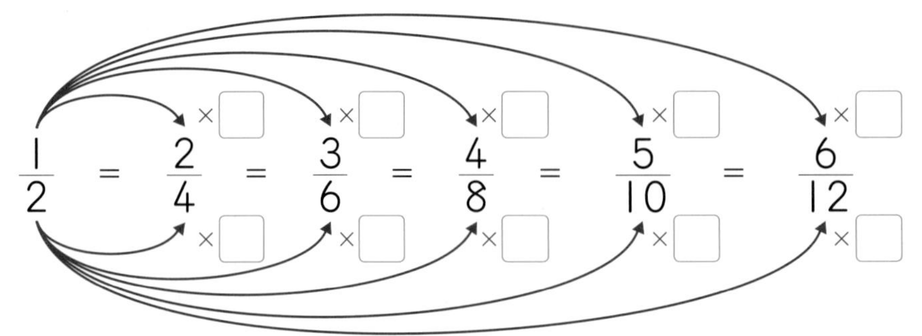


③ In opposite direction from ②, let's think a method to make $\frac{6}{12}$, $\frac{5}{10}$, $\frac{4}{8}$, $\frac{3}{6}$, and $\frac{2}{4}$ into $\frac{1}{2}$.

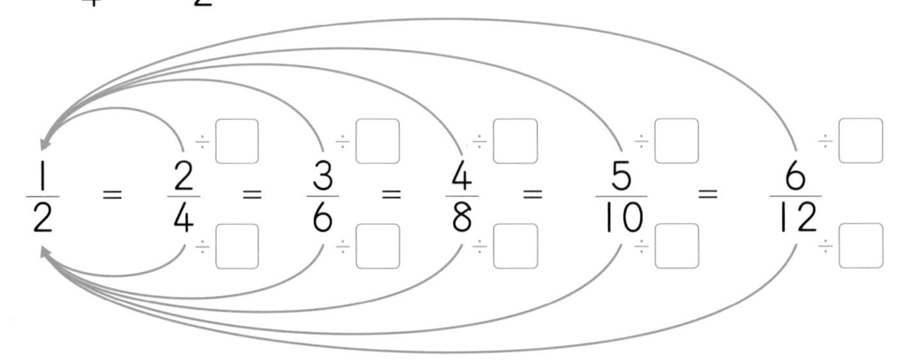


**Summary**

The size of a fraction does not change even if the numerator and denominator are multiplied or divided by the same number.

$$\frac{▲}{●} = \frac{▲ \times ■}{● \times ■}, \quad \frac{▲}{●} = \frac{▲ \div ■}{● \div ■}$$

**Want to confirm**

**1** Each of the following numerators and denominators is how many times of $\frac{1}{3}$?

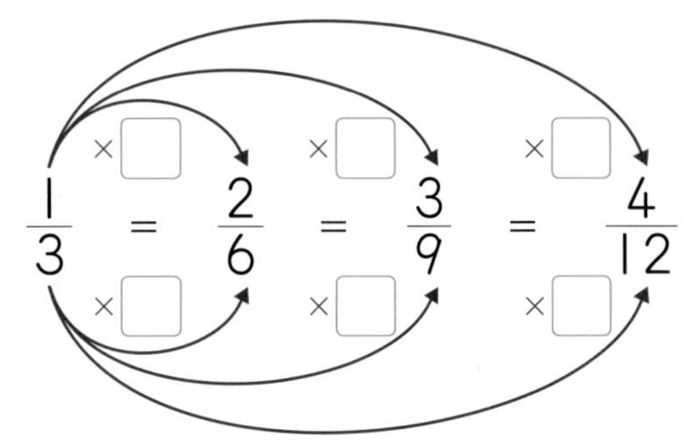

**2** What should we do to make $\frac{2}{10}$, $\frac{3}{15}$, and $\frac{4}{20}$ into $\frac{1}{5}$?

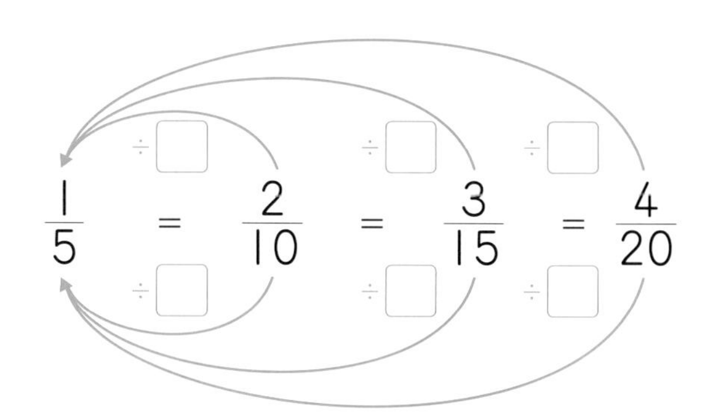


Want to try

**3** Let's write three equivalent fractions to the following fractions.

① $\frac{1}{4} = \frac{2}{\square} = \frac{\square}{12} = \frac{4}{\square}$　　② $\frac{8}{24} = \frac{4}{\square} = \frac{\square}{6} = \frac{1}{\square}$

## That's it

### How many unit fractions?

We can express a fraction into various fractions having the same size by changing its denominator. Let's represent $\frac{3}{4}$ into different fractions using 8 and 12 as denominators.

$$\frac{3}{4} = \frac{3 \times \square}{4 \times \square} = \frac{\square}{8}$$

$$\frac{3}{4} = \frac{3 \times \square}{4 \times \square} = \frac{\square}{12}$$

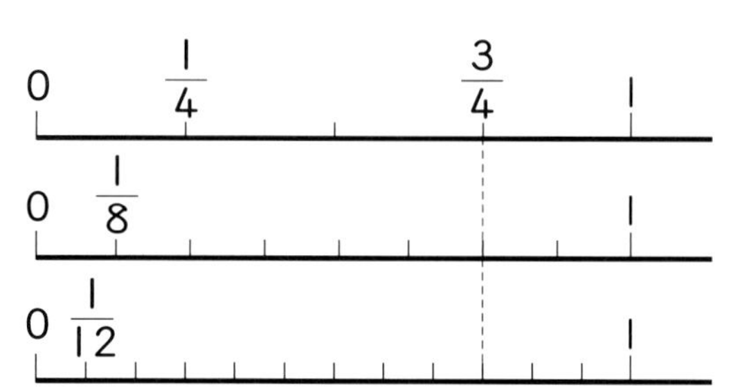


Fractions such as $\frac{1}{8}$ and $\frac{1}{12}$ are called **unit fractions**. We can find the size of fractions by counting its unit fractions.

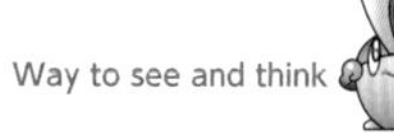


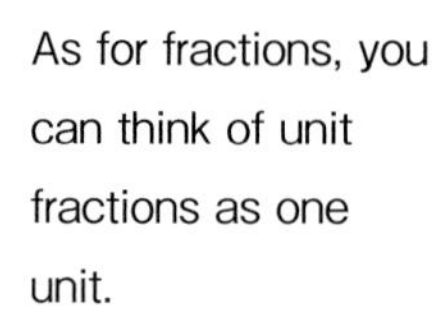

As for fractions, you can think of unit fractions as one unit.

How many unit fractions are there in the following fractions?

① $\frac{6}{8}$ has □ of the unit fraction $\frac{1}{8}$.

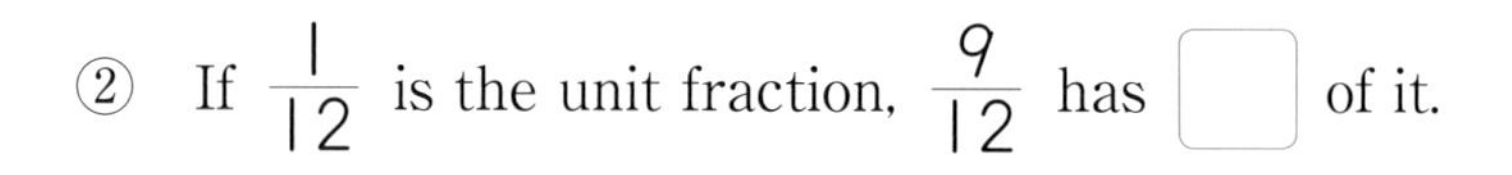

② If $\frac{1}{12}$ is the unit fraction, $\frac{9}{12}$ has □ of it.

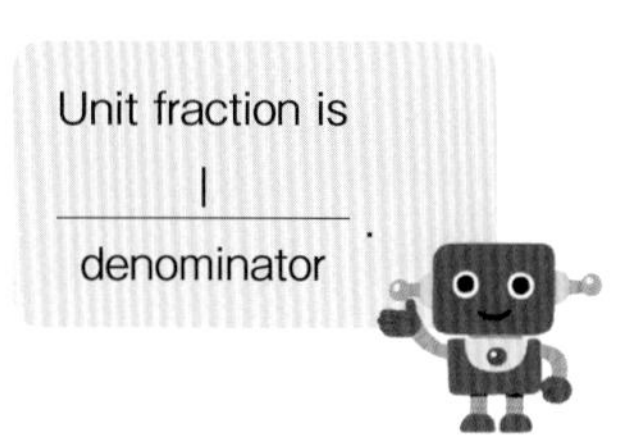

Want to know　Reducing fractions

## 2 Let's find the equivalent fraction to $\frac{12}{18}$ that has the smallest denominator.

Yui: To get an equivalent size, I should multiply or divide the denominator and numerator by the same number.

Hiroto: Since we want to make the denominator smaller, should we divide the denominator and numerator?

**Purpose** What should we do to reduce the numerator and denominator without changing the size of the fraction?

Want to explain

① Daiki thought as shown on the right.

Let's explain Daiki's idea.

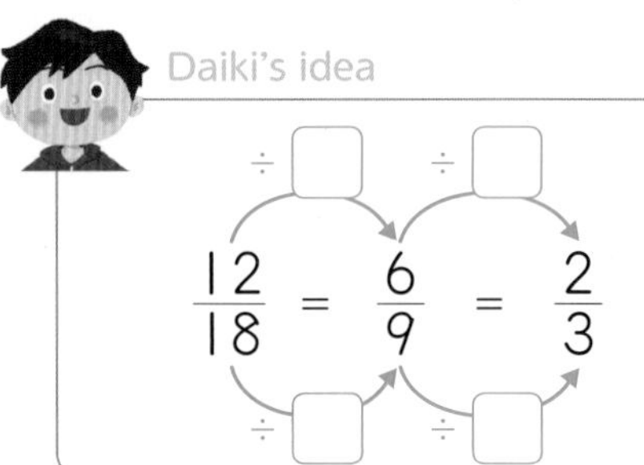


Dividing the numerator and denominator by a common divisor to make a simpler fraction is called **reducing fractions**. When reducing a fraction, we usually divide it until we get the smallest numerator and denominator.

② $\frac{12}{18}$ was reduced using the methods shown in Ⓐ and Ⓑ. Let's explain how the fraction was reduced.

Ⓐ $\frac{\not{12}^{\,\not{6}\,2}}{\not{18}_{\,\not{9}\,3}} = \frac{2}{3}$

Ⓑ $\frac{\not{12}^{\,2}}{\not{18}_{\,3}} = \frac{2}{3}$

**Summary**

We can reduce a fraction at once by using the greatest common divisor of the denominator and numerator.

Want to confirm

Let's reduce the following fractions.

① $\frac{8}{10}$　② $\frac{18}{27}$　③ $\frac{16}{40}$　④ $2\frac{6}{9}$

Want to know　Common denominators

## 3

**Which one is larger, $\frac{2}{3}$ or $\frac{4}{5}$? Let's compare the sizes.**

Daiki

Can we compare even if the denominators are different?

What should I do to get equal denominators?

Nanami

**Purpose** What should we do to compare fractions with different denominators?

① Let's make equivalent fractions to $\frac{2}{3}$ and $\frac{4}{5}$, respectively.

| $\frac{2}{3}$ | $\frac{4}{6}$ | $\frac{\square}{9}$ | $\frac{8}{12}$ | $\frac{\square}{15}$ | $\frac{\square}{18}$ | $\frac{14}{21}$ | $\frac{16}{24}$ | $\frac{18}{27}$ | $\frac{\square}{30}$ | … |
|---|---|---|---|---|---|---|---|---|---|---|
| $\frac{4}{5}$ | $\frac{8}{10}$ | $\frac{\square}{15}$ | $\frac{\square}{20}$ | $\frac{20}{25}$ | $\frac{\square}{30}$ | $\frac{28}{35}$ | $\frac{\square}{40}$ | $\frac{36}{45}$ | $\frac{\square}{50}$ | … |

② Which fractions can be compared? Let's write a ◯ in the diagram from ①.
Also, what kind of number is the denominator of the compared fractions?

> Changing fractions with different denominators into fractions with the same denominator without changing the sizes is called **changing fractions to common denominators**.

Way to see and think
When changing to common denominators, both numbers are aligned to the same unit fraction.

Based on the denominator of each original fraction, the common denominators 15, 30, ..., are common multiples of 3 and 5.

**Summary**

> We can compare fractions with different denominators by changing them into fractions with common denominators.

Want to confirm

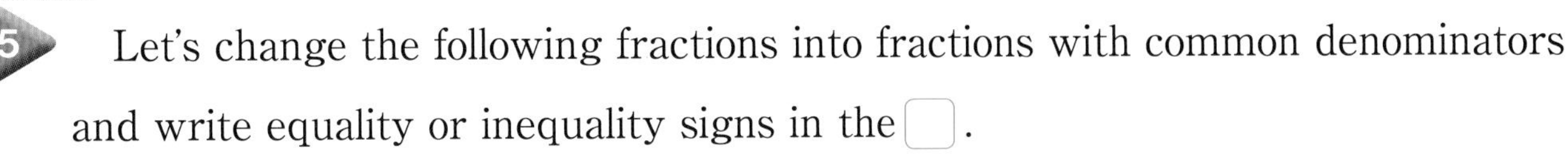

**5** Let's change the following fractions into fractions with common denominators and write equality or inequality signs in the $\square$.

① $\frac{3}{4}\ \square\ \frac{4}{5}$　② $\frac{1}{2}\ \square\ \frac{3}{8}$　③ $\frac{5}{6}\ \square\ \frac{8}{9}$　④ $\frac{6}{12}\ \square\ \frac{4}{8}$

Want to know How to find common denominators

Activity

**4** **The common denominators of $\frac{5}{6}$ and $\frac{7}{8}$ were found as follows. Let's explain the ideas of the children.**

Yui's idea

$$\frac{5}{6}=\frac{5\times8}{6\times8}=\frac{40}{48}$$
$$\frac{7}{8}=\frac{7\times6}{8\times6}=\frac{42}{48}$$

Hiroto's idea

$$\frac{5}{6}=\frac{5\times4}{6\times4}=\frac{20}{24}$$
$$\frac{7}{8}=\frac{7\times3}{8\times3}=\frac{21}{24}$$

When changing to common denominators, usually we use the least common multiple as the denominator to make it smaller.

Want to confirm

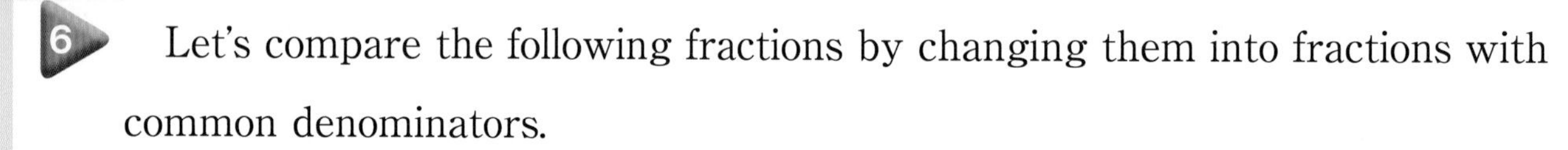

**6** Let's compare the following fractions by changing them into fractions with common denominators.

① $\frac{1}{4}$ and $\frac{2}{7}$

The least common multipe of 4 and 7 is □.

$\frac{1}{4}=\frac{1\times\square}{4\times\square}=\frac{\square}{\square}$, $\frac{2}{7}=\frac{2\times\square}{7\times\square}=\frac{\square}{\square}$. Thus, $\frac{1}{4}\ \square\ \frac{2}{7}$.

② $\frac{1}{3}$ and $\frac{2}{9}$

The least common multiple of 3 and 9 is □.

$\frac{1}{3}=\frac{1\times\square}{3\times\square}=\frac{\square}{\square}$. Thus, $\frac{1}{3}\ \square\ \frac{2}{9}$.

Want to try

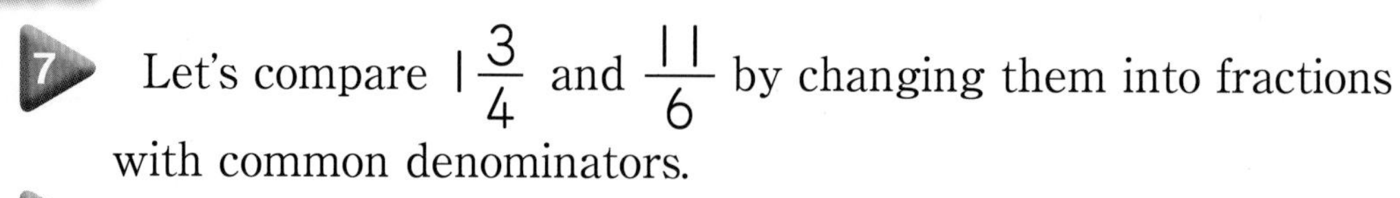

**7** Let's compare $1\frac{3}{4}$ and $\frac{11}{6}$ by changing them into fractions with common denominators.

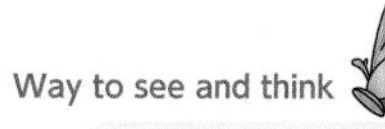

Way to see and think

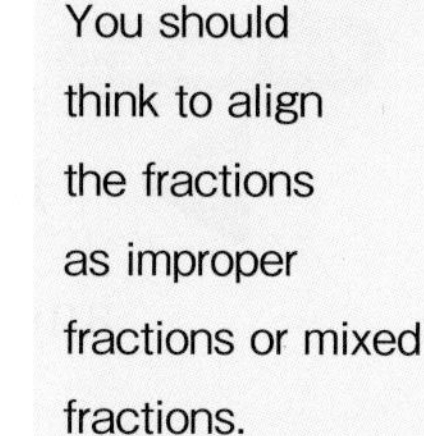

You should think to align the fractions as improper fractions or mixed fractions.

**8** Let's compare the following fractions by changing them into fractions with common denominators.

① $\left(\frac{2}{3}, \frac{3}{4}, \frac{4}{5}\right)$ ② $\left(\frac{3}{8}, \frac{2}{6}, \frac{1}{4}\right)$

## Addition of fractions

Want to think | Addition of proper fractions

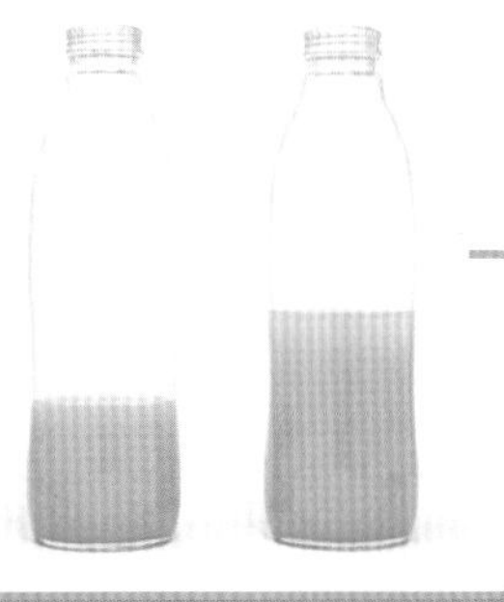

**1** **Two bottles contain $\frac{1}{3}$ L and $\frac{1}{2}$ L of juice.**

**How many liters are there altogether?**

The denominators are different.

Since fractions with the same denominator can be calculated...

**Purpose** What should we do for the addition of fractions with different denominators?

① Let's write a math expression.

Want to explain

② Let's explain the following calculation method.

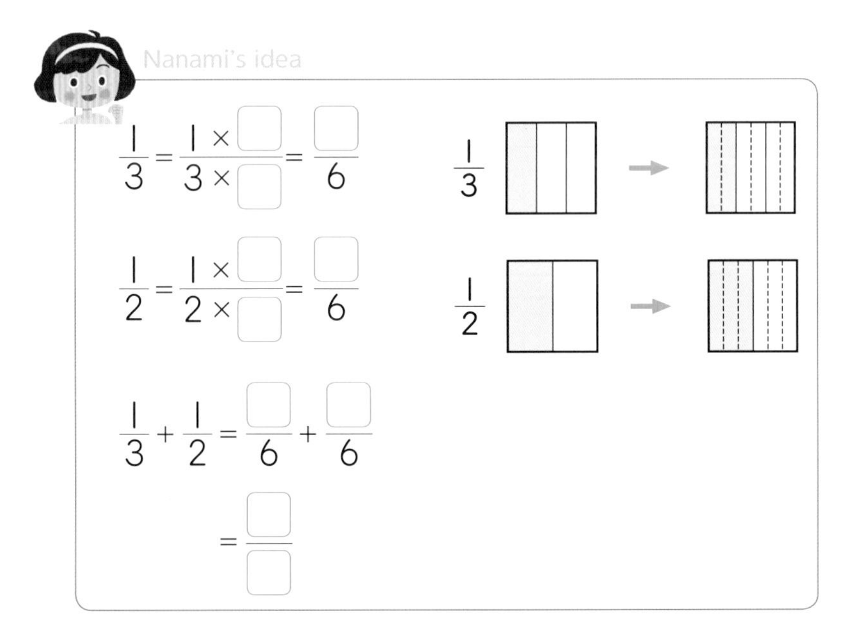


Way to see and think

You should think of the number of units when the fractions are aligned by unit fractions.

**Summary**

We can add fractions with different denominators by changing them into fractions with common denominators.

Want to think

1 Let's think about how to solve the following calculations.

① $\frac{3}{10}+\frac{1}{6}=\frac{\square}{\square}+\frac{\square}{\square}$
$=\frac{\square}{\square}$
$=\square$

② $\frac{3}{4}+\frac{1}{12}=\frac{\square}{\square}+\frac{\square}{\square}$
$=\frac{\square}{\square}$
$=\square$

If the answer can be reduced, we should reduce it to its simplest form.

2 Let's think about how to calculate $\frac{1}{3}+\frac{5}{6}$.

$\frac{1}{3}+\frac{5}{6}=\frac{\square}{\square}+\frac{\square}{6}$
$=\frac{\square}{\square}$
$=\square$

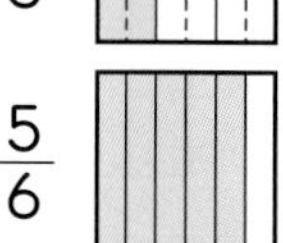


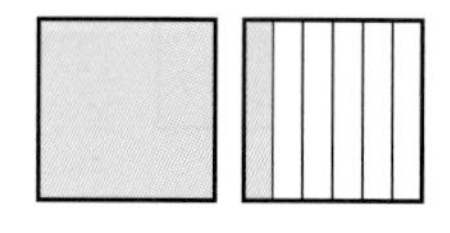

When the answer becomes an improper fraction, it is easier to understand the size by changing it to a mixed fraction.

Daiki

Want to confirm

3 Let's solve the following calculations.

① $\frac{1}{2}+\frac{1}{5}$ ② $\frac{1}{6}+\frac{2}{7}$ ③ $\frac{3}{7}+\frac{1}{2}$

④ $\frac{1}{12}+\frac{5}{8}$ ⑤ $\frac{1}{6}+\frac{7}{10}$ ⑥ $\frac{3}{8}+\frac{7}{10}$

⑦ $\frac{5}{12}+\frac{1}{3}$ ⑧ $\frac{4}{5}+\frac{13}{15}$ ⑨ $\frac{11}{12}+\frac{1}{4}$

Want to think Addition of mixed fractions

## 2 Goods with various weights will be placed inside a box that weighs $1\frac{1}{2}$ kg. Let's think about the total weight at this time.

① Let's think about how to calculate when the weight of the goods is $1\frac{1}{6}$ kg.

$$1\frac{1}{2} + 1\frac{1}{6} = 1\frac{\square}{\square} + 1\frac{1}{6}$$

$$= \square\frac{\square}{\square}$$

$$= \square$$

$1\frac{1}{2}$

$+\ 1\frac{1}{6}$

② Let's think about how to calculate when the weight of the goods is $1\frac{2}{3}$ kg.

$$1\frac{1}{2} + 1\frac{2}{3} = 1\frac{\square}{6} + 1\frac{\square}{6}$$

$$= \square\frac{\square}{6}$$

$$= \square\frac{\square}{6}$$

$1\frac{1}{2}$

$+\ 1\frac{2}{3}$

③ For problems ① and ②, Nanami changed the mixed fractions into improper fractions. Let's calculate by using Nanami's way of thinking.

Want to confirm

Let's solve the following calculations.

① $2\frac{1}{8} + 1\frac{3}{4}$ ② $1\frac{2}{3} + 3\frac{1}{4}$ ③ $\frac{1}{6} + 2\frac{8}{9}$

④ $1\frac{7}{15} + 1\frac{7}{10}$ ⑤ $1\frac{5}{6} + 1\frac{1}{2}$ ⑥ $2\frac{2}{3} + 1\frac{7}{12}$

## 3 Subtraction of fractions

Want to think — Subtraction of proper fractions

**1** **There are $\frac{3}{4}$ L of juice and $\frac{5}{8}$ L of milk. How many liters is the difference in quantity?**

Yui

Since the denominators are different, I don't know which one is bigger.

The same as with the addition, we should first change to common denominators.

Hiroto

**Purpose** What should we do for the subtraction of fractions with different denominators?

① Let's change to fractions with common denominators. After exploring which is larger, let's write a math expression.

$\frac{3}{4} = \frac{\square}{\square}$, thus $\frac{3}{4} \square \frac{5}{8}$.

Math expression: $\square$

Way to see and think

The same as with addition, you can compare sizes by changing to common denominators.

Want to explain

② Let's explain the following calculation method.

Nanami's idea

$\frac{3}{4} = \frac{3 \times \square}{4 \times \square} = \frac{\square}{8}$

$\frac{3}{4} - \frac{5}{8} = \frac{\square}{8} - \frac{\square}{8}$

$= \frac{\square}{\square}$

$\frac{3}{4}$

$- \frac{5}{8}$

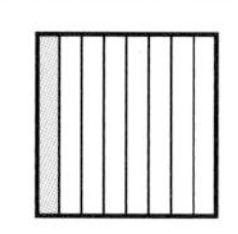

**Summary**

We can subtract fractions with different denominators by changing them into fractions with common denominators.

Want to confirm

Let's solve the following calculations.

① $\frac{6}{7} - \frac{3}{4}$ ② $\frac{3}{4} - \frac{7}{10}$ ③ $\frac{7}{9} - \frac{1}{6}$ ④ $\frac{5}{8} - \frac{1}{4}$

Want to try

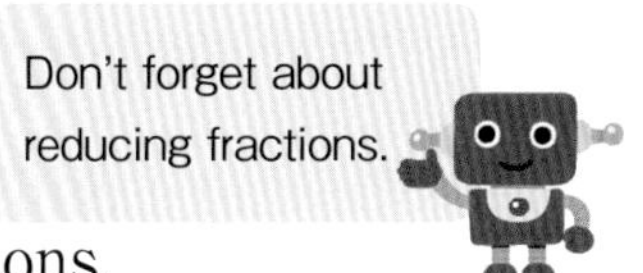
Don't forget about
reducing fractions.

**2** Let's think about how to solve the following calculations.

① $\frac{5}{6}-\frac{3}{10}=\frac{\square}{\square}-\frac{\square}{\square}$

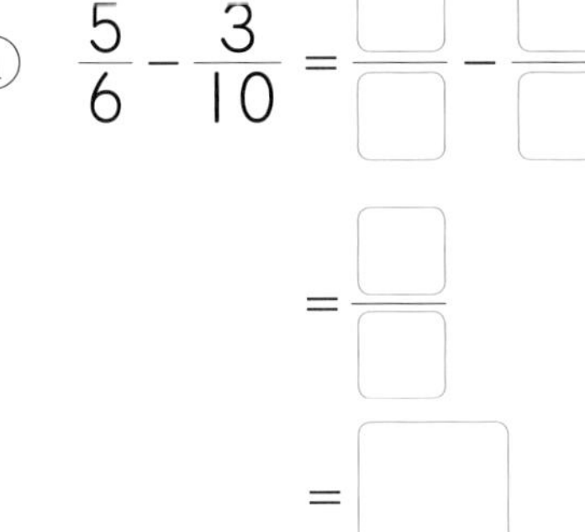

$=\frac{\square}{\square}$

$=\square$

② $\frac{7}{5}-\frac{5}{6}=\frac{\square}{\square}-\frac{\square}{\square}$

$=\square$

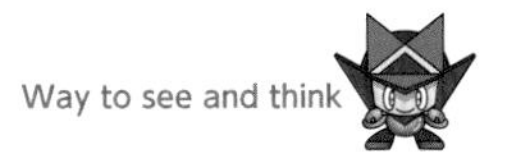
Way to see and think

You can calculate improper fraction − proper fraction in the same way by changing to common denominators.

Want to confirm

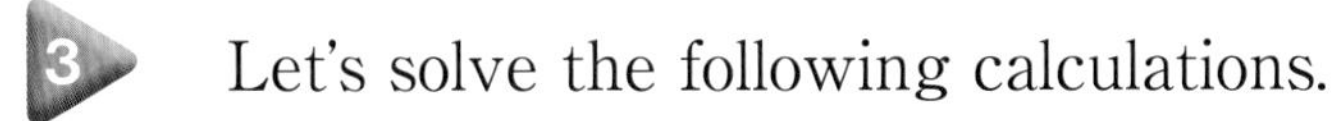

**3** Let's solve the following calculations.

① $\frac{2}{3}-\frac{1}{6}$ ② $\frac{2}{5}-\frac{1}{15}$ ③ $\frac{7}{15}-\frac{3}{10}$ ④ $\frac{13}{11}-\frac{2}{3}$

⑤ $\frac{9}{8}-\frac{5}{6}$ ⑥ $\frac{5}{4}-\frac{9}{20}$ ⑦ $\frac{11}{10}-\frac{4}{15}$ ⑧ $\frac{6}{5}-\frac{10}{9}$

Want to think Subtraction of mixed fractions

Activity

**2** **There were $2\frac{1}{2}$ L of juice at the house. Let's find the remaining amount of juice when some liters were drunk in a week.**

① Let's think about how to calculate when $1\frac{1}{6}$ L were drunk in a week.

$2\frac{1}{2}-1\frac{1}{6}=2\frac{\square}{\square}-1\frac{1}{6}$

$=\square\frac{\square}{\square}$

$=\square$

$2\frac{1}{2}$

$-\ 1\frac{1}{6}$

② Let's think about how to calculate when $1\frac{5}{6}$ L were drunk in a week.

Yui's idea

Change the mixed fractions into improper fractions. $2\frac{1}{2}=\frac{\square}{2}$, $1\frac{5}{6}=\frac{\square}{6}$

So, $2\frac{1}{2}-1\frac{5}{6}=\frac{\square}{2}-\frac{\square}{6}=\frac{\square}{6}-\frac{\square}{6}=\frac{\square}{6}$

Reducing it, $\frac{\square}{6}=\frac{\square}{\square}$

Hiroto's idea

Subtract whole numbers and proper fractions separately.

$2\frac{1}{2}-1\frac{5}{6}=2\frac{3}{6}-1\frac{5}{6}$

Since we cannot subtract $\frac{5}{6}$ from $\frac{3}{6}$,

borrow 1 from 2. $2\frac{3}{6}=1\frac{9}{6}$

$1\frac{9}{6}-1\frac{5}{6}=\frac{\square}{6}=\frac{\square}{\square}$

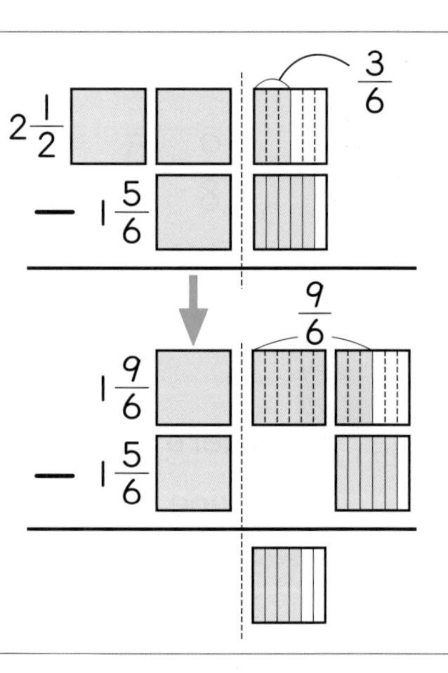

**4** Let's solve the following calculations.

① $7\frac{3}{4}-2\frac{1}{6}$ ② $4\frac{7}{8}-1\frac{1}{7}$ ③ $3\frac{5}{9}-\frac{2}{7}$

④ $5\frac{2}{3}-2\frac{1}{6}$ ⑤ $7\frac{2}{5}-4\frac{5}{7}$ ⑥ $1\frac{2}{9}-\frac{5}{6}$

⑦ $2\frac{2}{5}-\frac{7}{10}$ ⑧ $5\frac{1}{6}-3\frac{9}{10}$ ⑨ $7\frac{1}{4}-6\frac{11}{12}$

Want to know Calculation that includes addition and subtraction

Activity

**3** **Let's think about how to calculate $\frac{1}{2}+\frac{2}{3}-\frac{1}{4}$.**

Want to compare

① Let's compare the ways of thinking of the following children.

Hiroto's idea

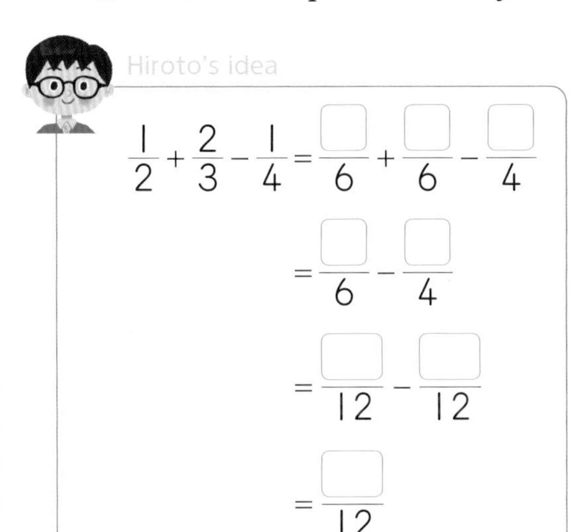

$$\frac{1}{2}+\frac{2}{3}-\frac{1}{4}=\frac{\square}{6}+\frac{\square}{6}-\frac{\square}{4}$$
$$=\frac{\square}{6}-\frac{\square}{4}$$
$$=\frac{\square}{12}-\frac{\square}{12}$$
$$=\frac{\square}{12}$$

Yui's idea

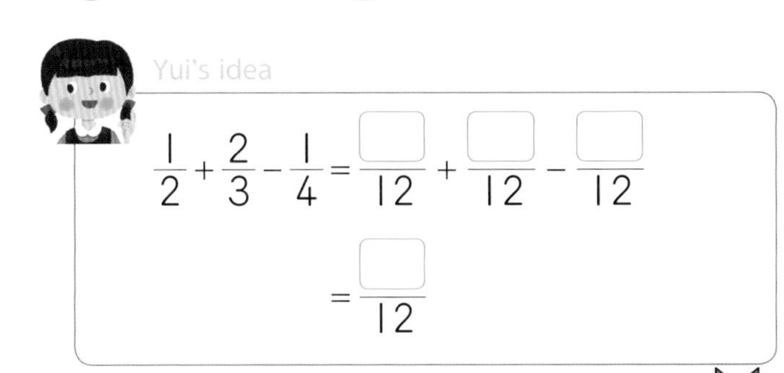

$$\frac{1}{2}+\frac{2}{3}-\frac{1}{4}=\frac{\square}{12}+\frac{\square}{12}-\frac{\square}{12}$$
$$=\frac{\square}{12}$$

Way to see and think

Even if the number of additions or subtractions increases, you can calculate in the same way.

Want to confirm

**5** Let's solve the following calculations.

① $\frac{1}{4}+\frac{3}{8}-\frac{1}{2}$ ② $\frac{5}{4}+\frac{4}{3}-\frac{5}{6}$

③ $\frac{2}{3}-\frac{5}{12}+\frac{3}{4}$ ④ $\frac{4}{5}-\frac{3}{10}-\frac{1}{2}$

Want to try

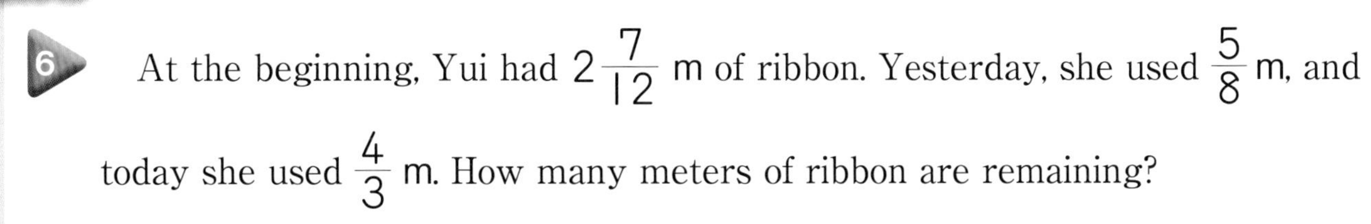

**6** At the beginning, Yui had $2\frac{7}{12}$ m of ribbon. Yesterday, she used $\frac{5}{8}$ m, and today she used $\frac{4}{3}$ m. How many meters of ribbon are remaining?

# What you can do now

☐ **Understanding how to reduce fractions.**

**1** Let's reduce the following fractions and find the fractions that are equivalent to $\frac{3}{4}$.

Ⓐ $\frac{6}{8}$ Ⓑ $\frac{8}{12}$ Ⓒ $\frac{16}{20}$ Ⓓ $\frac{24}{32}$ Ⓔ $\frac{30}{40}$

☐ **Understanding how to change fractions into fractions with common denominators.**

**2** Let's change the following fractions into fractions with common denominators and write inequality signs in the ☐.

① $\frac{2}{3}$ ☐ $\frac{1}{2}$ ② $\frac{3}{4}$ ☐ $\frac{5}{7}$ ③ $\frac{1}{6}$ ☐ $\frac{5}{18}$ ④ $\frac{4}{9}$ ☐ $\frac{5}{12}$

☐ **Can add and subtract fractions.**

**3** Let's solve the following calculations.

① $\frac{1}{6}+\frac{2}{5}$ ② $\frac{2}{7}+\frac{1}{4}$ ③ $\frac{1}{12}+\frac{2}{3}$ ④ $\frac{5}{6}+\frac{2}{3}$

⑤ $\frac{3}{5}+\frac{4}{7}$ ⑥ $\frac{1}{4}+\frac{5}{6}$ ⑦ $1\frac{1}{2}+1\frac{9}{10}$ ⑧ $1\frac{5}{6}+2\frac{4}{9}$

⑨ $1\frac{2}{3}+1\frac{8}{15}$ ⑩ $2\frac{5}{6}+4\frac{5}{12}$ ⑪ $\frac{3}{7}-\frac{2}{5}$ ⑫ $\frac{7}{9}-\frac{1}{6}$

⑬ $\frac{4}{9}-\frac{5}{18}$ ⑭ $\frac{7}{8}-\frac{5}{24}$ ⑮ $1\frac{1}{3}-\frac{1}{4}$ ⑯ $6\frac{5}{7}-2\frac{3}{5}$

⑰ $3\frac{3}{4}-1\frac{5}{6}$ ⑱ $1\frac{1}{7}-\frac{3}{4}$ ⑲ $2\frac{2}{3}-1\frac{1}{6}$ ⑳ $3\frac{1}{6}-1\frac{3}{4}$

☐ **Can solve problems by using the addition and subtraction of fractions.**

**4** Hayato has $\frac{3}{4}$ m of ribbon. Yuka has $\frac{4}{5}$ m of ribbon. Let's answer the following questions.

① Which ribbon is longer and by how many meters?

② If you connect the two ribbons together, how many meters is the total length?

Supplementary Problems p.150

# Usefulness and efficiency of learning

**1** Let's reduce the following fractions.

① $\frac{49}{63}$　　② $\frac{30}{42}$　　③ $\frac{45}{100}$

☐ Understanding how to reduce fractions.

**2** Is the following calculation correct? If it is wrong, let's explain the reasons.

$\frac{1}{3}+\frac{2}{5}=\frac{3}{8}$

☐ Understanding how to change fractions into fractions with common denominators.

**3** There are $\frac{3}{4}$ L of milk coffee and $\frac{5}{6}$ L of milk. Let's answer the following questions.

① Which is more and by how many liters?

② How many liters are there altogether?

☐ Can solve problems by using the addition and subtraction of fractions.

**4** Sota is going to a river to fish. Now, he is $4\frac{1}{2}$ km from his house, and he has to go $\frac{5}{8}$ km more to reach the river. How many kilometers is the distance from his house to the river?

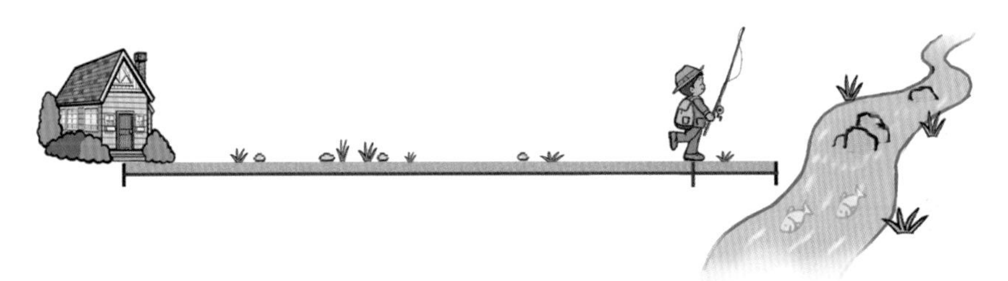

☐ Can solve problems by using the addition and subtraction of fractions.

**5** There is a $\frac{7}{10}$ kg basket that weighs $3\frac{1}{4}$ kg when apples are placed inside it. How many kilograms is the weight of the apples?

☐ Can solve problems by using the addition and subtraction of fractions.

# How many liters for one person?

**Problem** As for an indivisible amount, how should it be represented?

Fractions, Decimal Numbers, and Whole Numbers

# Let's think about the relationship among fractions, decimal numbers and whole numbers, and their structure.

## 1 Quotients and fractions

Want to know

**1** **When you equally divide 2 L of juice among □ children, how many liters of juice will each child receive?**

① Let's write the whole number from 1 to 5 in each □ and solve the calculations.

| | | | |
|---|---|---|---|
| 1 child | 2 ÷ □ = | 4 children | 2 ÷ □ = |
| 2 children | 2 ÷ □ = | 5 children | 2 ÷ □ = |
| 3 children | 2 ÷ □ = | | |

Want to categorize

② Classify the math expressions in ① into 3 groups based on the answers.

Ⓐ Answer is a whole number. (          )

Ⓑ Answer is expressed exactly as a decimal number. (          )

Ⓒ Answer is indivisible. (          )

2 ÷ 3 is 0.666... and there is no end, so the answer is indivisible.

Daiki: We cannot calculate answers that are indivisible?

Nanami: But, when we learned fractions, there was a way to represent it as $\frac{1}{3}$ ...

**Purpose** How should we represent the quotient of an indivisible division?

③ When you equally divide 2 L of juice among 3 children, how many liters of juice will each child receive? Let's color the amount of juice for one person.

1L 1L

④ How many liters of juice will each child receive? Let's think by looking at the diagrams.

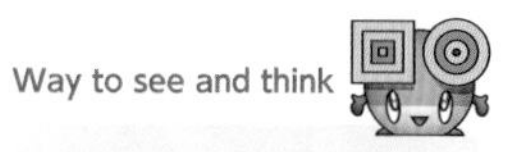


You can think of $\frac{2}{3}$ as 2 sets of $\frac{1}{3}$.

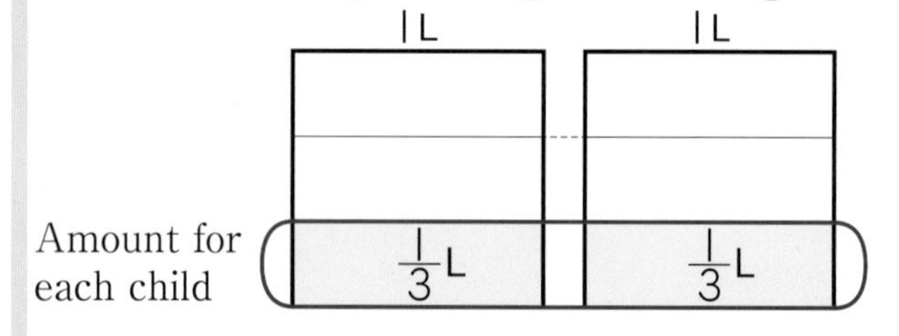


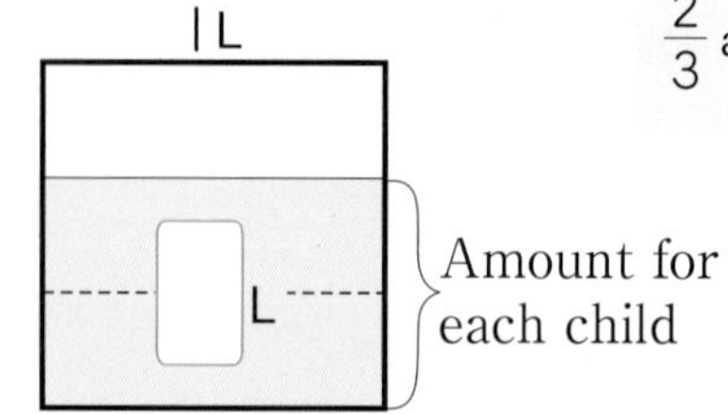


When 1L is equally divided into 3 parts, the amount for one child is ☐ L, so when 2L are equally divided into 3 parts, two sets of the part are ☐ L.

→ $2 \div 3 = \frac{☐}{☐}$

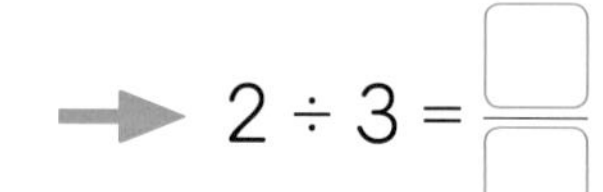

Want to confirm

**1** If you divide a 3 m string into 4 equal parts, how many meters is the length of one part?

① Let's write a math expression. ☐

② Let's find the length of one part.

1 m into 4 equal parts → $1 \div 4 =$ ☐

2 m into 4 equal parts → $2 \div 4 =$ ☐

3 m into 4 equal parts → $3 \div 4 =$ ☐

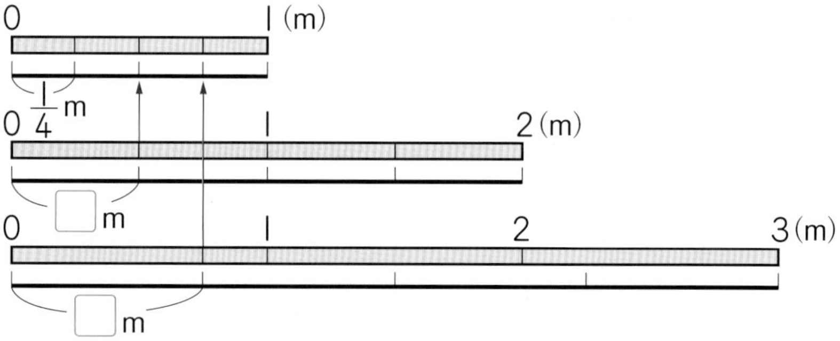

(Number lines: 0 to 1 (m), with $\frac{1}{4}$ m; 0 to 2 (m), with ☐ m; 0 to 3 (m), with ☐ m)

**Summary**

The quotient of a division problem in which a whole number is divided by another whole number can be represented as a fraction.

$● \div ▲ = \frac{●}{▲}$

Want to try

**2** Let's represent the following quotients as fractions.

① $1 \div 6$ ② $5 \div 8$ ③ $8 \div 3$ ④ $9 \div 7$

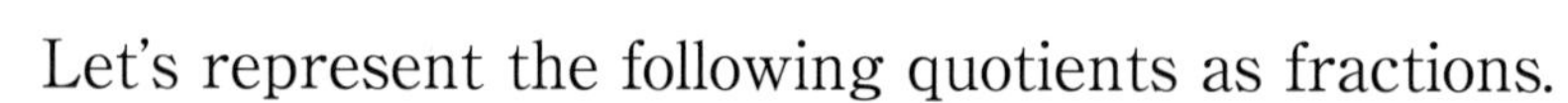

**3** Let's write the numbers that apply in the following ☐.

① $4 \div ☐ = \frac{4}{9}$ ② $☐ \div 5 = \frac{1}{5}$ ③ $☐ \div ☐ = \frac{7}{3}$

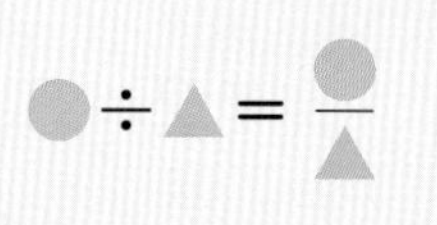

On exercise ③ from 3, we can place various numbers.

## 2

**There are ribbons as shown below. How many times of the length of the white ribbon is the length of the red and blue ribbons?**

Length of ribbons

| Color | Length (m) |
|---|---|
| Red | 4 |
| White | 3 |
| Blue | 2 |

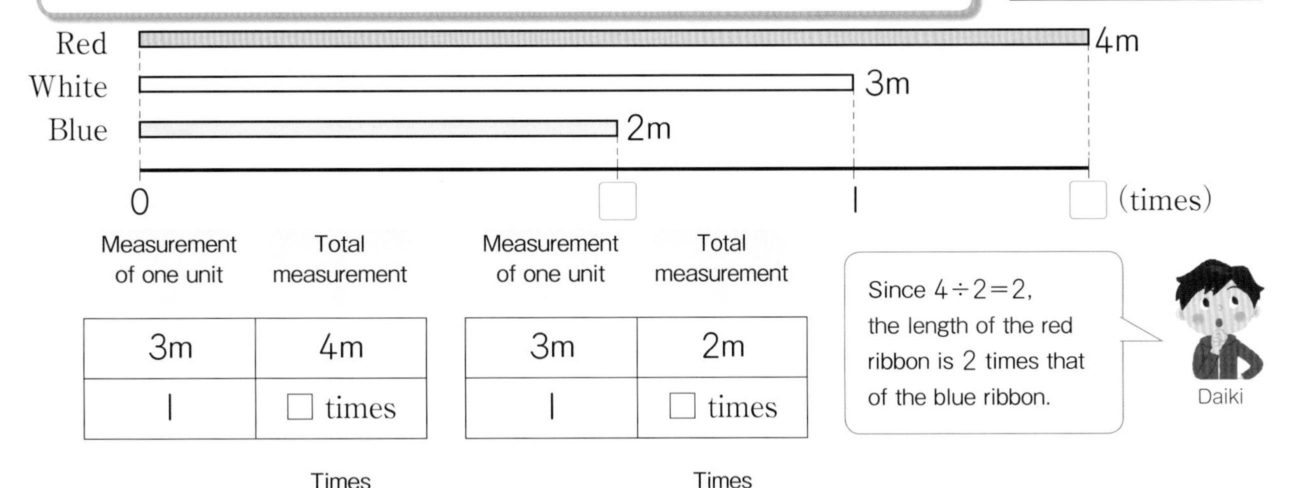


① How many times of the length of the white ribbon is the length of the red ribbon?

$4 \div 3 = \frac{\square}{\square}$ $\square$ times

② How many times of the length of the white ribbon is the length of the blue ribbon? Let's write a math expression and find the answer.

$\square$ $\square$ times

We can use fractions to represent how many times of a number such as $\frac{4}{3}$ times or $\frac{2}{3}$ times.

Want to confirm

There are 7 L of water in a tank and 3 L of water in a bucket. Let's answer the following questions.

① How many times of the water in the tank is the water in the bucket?

② How many times of the water in the bucket is the water in the tank?

## 2 Fractions, decimal numbers, and whole numbers

Want to represent Changing fractions to decimal numbers or whole numbers

**1** **If we divide a 2m tape into 5 equal parts, how many meters is the length of each part?**

Since we divide 2m into 5 equal parts, it can be found with 2÷5.

It looks like we can represent it by a fraction and decimal number.

Purpose What kind of relationship is there between fractions and decimal numbers or whole numbers?

① Let's represent the answer as a fraction and decimal number.

$2 \div 5 = \frac{\square}{\square}$ $\qquad$ $2 \div 5 = \square$

② Let's represent the fraction and decimal number from ① in the following number line.

Way to see and think

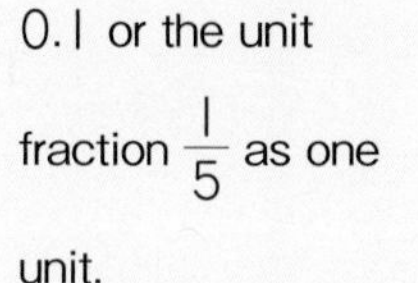

You can think of 0.1 or the unit fraction $\frac{1}{5}$ as one unit.


0 0.1 $\qquad$ 1 $\qquad$ 2 (m)

0 $\frac{1}{5}$ $\qquad$ 1 $\qquad$ 2 (m)

$\frac{2}{5} = \square$

③ Let's represent the length of one part when a 4 m tape is divided into 2 equal parts as a fraction and whole number.

Summary

If we divide the numerator by the denominator, we can represent a fraction as a decimal number or whole number.

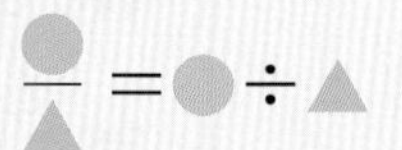

Want to confirm

**1** Let's represent the following fractions as decimal numbers or whole numbers.

① $\frac{7}{10} = \square$ $\qquad$ ② $\frac{29}{100} = \square$

③ $\frac{12}{4} = 12 \div 4 = \square$ $\qquad$ ④ $1\frac{3}{5} = \frac{8}{5} = 8 \div 5 = \square$

**Let's represent the following decimal numbers as fractions.**

① 0.3 ② 1.47

I can change a fraction to a decimal number, but could I do the reverse?

Since 0.1 is $\frac{1}{10}$ and 0.01 is $\frac{1}{100}$, …

**Purpose** Can decimal numbers be represented as fractions?

① 0.3 is $\square$ sets of 0.1.

↓

Since there are $\square$ sets of $\frac{1}{10}$, then $\frac{\square}{\square}$.

② 1.47 is $\square$ sets of 0.01.

↓

Since there are $\square$ sets of $\frac{1}{100}$, then $\frac{\square}{\square}$.

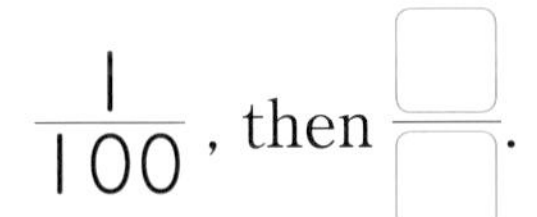

**Summary**

Decimal numbers can be represented as fractions if we choose $\frac{1}{10}$ or $\frac{1}{100}$ as the units.

The following whole numbers are represented as fractions. Let's write the numbers that apply in the $\square$.

① 2

$2 = 2 \div 1 = \frac{2}{1}$

$2 = 4 \div 2 = \frac{4}{2}$

$2 = 8 \div \square = \square$

② 5

$5 = 5 \div 1 = \square$

$5 = 10 \div 2 = \square$

$5 = 30 \div \square = \square$

Whole numbers can be represented as fractions no matter what whole number we choose for the denominator.

Want to confirm

**3** Let's change the following decimal numbers or whole numbers into fractions.

① 0.7 ② 0.67 ③ 4 ④ 12 ⑤ 3.14

Want to try

**4** Let's find the decimal numbers or fractions that apply in the following □.

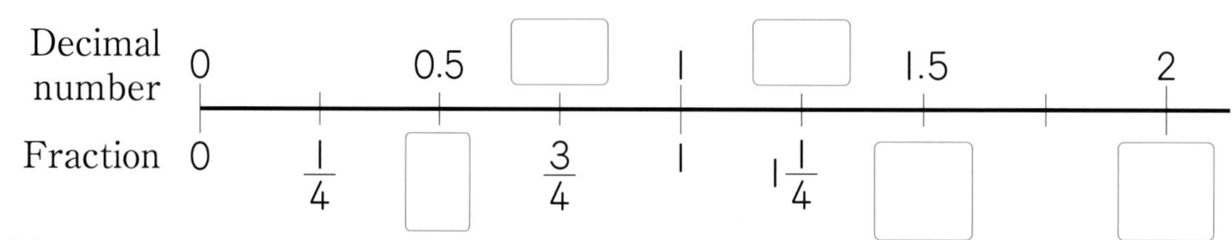


Want to confirm

**5** Let's change the following decimal numbers into fractions and the fractions into decimal numbers or whole numbers.

① 0.9 ② 1.05 ③ $\frac{3}{5}$ ④ $\frac{24}{6}$ ⑤ $1\frac{2}{5}$

That's it — Develop in Junior High School

## Decimal number that goes on with the same number

If the following fractions are represented as decimal numbers, both become decimal numbers that go on with the same number after the decimal point.

$\frac{1}{3} = 0.333...$ $\frac{1}{9} = 0.111...$

If you use these numbers, you can represent decimal numbers that go on with the same number as fractions.

$$0.22... = 0.11... + 0.11... = \frac{1}{9} + \frac{1}{9} = \frac{2}{9}$$

$$0.44... = 0.33... + 0.11... = \frac{1}{3} + \frac{1}{9} = \frac{3}{9} + \frac{1}{9} = \frac{4}{9}$$

Hiroto

Let's try to think if we can represent in the same way the following numbers: 0.55..., 0.66..., 0.77..., 0.88..., 0.99..., as fractions.

Want to know Comparing the size of fractions, decimal numbers, and whole numbers

**3** **Let's represent the following numbers by using ↓ on the number line below.**

$\frac{4}{5}$ 0.6 $1\frac{7}{20}$ 2 1.25

0 1

Whole numbers, decimal numbers, and fractions can be represented on the same number line. This makes a comparison of numbers easier.

Want to compare

From $\frac{2}{3}$, $\frac{13}{20}$, and 0.61, which is the largest number?

Nanami's idea

I thought by aligning as fractions.

$\frac{2}{3} = \frac{200}{300}$

$\frac{13}{20} = \frac{195}{300}$

$0.61 = \frac{61}{100} = \frac{183}{300}$

Therefore, $\frac{2}{3}$ is the largest.

Daiki's idea

I thought by aligning as decimal numbers.

$\frac{2}{3} = 0.66...$

$\frac{13}{20} = 0.65$

0.61

Therefore, $\frac{2}{3}$ is the largest.

Changing fractions to decimal numbers makes comparison easier.

$\frac{2}{3} = 2 \div 3 = 0.666...$ → about 0.67

Want to confirm

Let's arrange the following numbers in ascending order.

1.3 0.75 $\frac{4}{2}$ $1\frac{1}{2}$ $\frac{7}{10}$ $\frac{5}{7}$

# What you can do now

☐ **Understanding the relationship between quotients and fractions.**

**1** Let's write the numbers that apply in the following ☐.

① $2 \div 7 = \frac{2}{\square}$　② $6 \div 9 = \frac{\square}{9} = \frac{\square}{\square}$　③ $20 \div 8 = \frac{20}{\square} = \frac{\square}{\square}$

④ $\square \div 6 = \frac{1}{6}$　⑤ $9 \div \square = \frac{9}{2}$　⑥ $\square \div 3 = \frac{14}{3}$

☐ **Understanding the meaning of number of times in fractions.**

**2** Let's answer the following questions.

① How many times of 3 L of water is 5 L of water?

② How many times of 5 L of water is 3 L of water?

☐ **Understanding the relationship between fractions, decimal numbers, and whole numbers.**

**3** Let's represent the following fractions as decimal numbers or whole numbers and the decimal numbers as fractions.

① $\frac{1}{2}$　② $\frac{31}{100}$　③ $\frac{16}{8}$　④ $1\frac{1}{4}$

⑤ 0.8　⑥ 1.9　⑦ 0.12　⑧ 1.11

☐ **Can compare the size of fractions, decimal numbers, and whole numbers.**

**4** Let's write equality or inequality signs in the following ☐.

① $\frac{3}{7}$ ☐ 0.4　② $\frac{5}{4}$ ☐ 1.3

③ 1.69 ☐ $\frac{12}{7}$　④ $1\frac{7}{8}$ ☐ 1.78

⑤ 1.3 ☐ $1\frac{3}{10}$　⑥ $2\frac{1}{9}$ ☐ 2.01

Supplementary Problems → p.151

# Usefulness and efficiency of learning

**1** Let's answer the following questions.

① Let's represent the following quotients as the simplest fraction as possible.

ⓐ $4 \div 8$　　ⓑ $12 \div 60$　　ⓒ $45 \div 100$

② As for the following fractions, what kind of quotient are they?

ⓐ $\frac{2}{5}$　　ⓑ $\frac{3}{7}$　　ⓒ $\frac{11}{18}$

□ Understanding the relationship between quotients and fractions.

**2** The distance from the house to the station is 9 km and the distance from the house to the library is 5 km.

Let's answer the following questions.

① How many times of the distance from the house to the station is the distance from the house to the library?

② How many times of the distance from the house to the library is the distance from the house to the station?

□ Understanding the meaning of number of times in fractions.

**3** Let's answer the following questions. Additionally, let's represent all answers in both fractions and decimal numbers.

① If you equally divide an 8 m ribbon among 5 children, how many meters will each child receive?

② 8 ornaments will be made from 3 kg of clay. How many kilograms of clay will be used to make one ornament?

③ If you equally divide 13 L of paint into 4 containers, how many liters of paint will be poured into each container?

□ Understanding the relationship between fractions, decimal numbers, and whole numbers.

**4** Let's represent the following numbers by using ↓ on the number line below.

$1$　　$1\frac{5}{20}$　　$0.7$　　$\frac{2}{5}$　　$1.8$　　$\frac{7}{5}$

0　　　　　　　　　　　　　　2

□ Can represent fractions and decimal numbers in the number line.

# Who scored the most?

I made a shooting record of everyone's shots.


Shooting record

| | | | | | | | | | | |
|---|---|---|---|---|---|---|---|---|---|---|
| Daiki | × | ○ | ○ | ○ | × | × | ○ | ○ | × | ○ |
| Yui | ○ | ○ | × | × | ○ | × | ○ | × | × | ○ |
| Hiroto | ○ | × | ○ | × | ○ | ○ | × | ○ | | |

○scored ×missed

This time, I got the most number of scores.

Daiki made a lot of shots and obtained a high number of scores.

I was able to make few shots, but scored more than half.

So, can we say who got the best result?


**Problem** How should we compare the shooting results?

Ratio (1)

# Let's explore how to compare the total and its parts.

## 1 Ratio

Want to know Shooting results

Activity

**1** **Daiki and friends played a basketball game. The table below shows the shooting records. Let's try to think how to compare the shooting results to find who had the best result.**

| | Number of shots | Number of scores |
|---|---|---|
| Daiki | 10 | 6 |
| Yui | 10 | 5 |
| Hiroto | 8 | 5 |

① Let's compare the results of Daiki and Yui.

② Let's compare the results of Yui and Hiroto.

③ Let's compare the results of Daiki and Hiroto.

It seems that we can compare when the number of shots is aligned.
We can compare when the number of scores is the same.
As for measures per unit quantity, we compared them per liter or per km.


Purpose How should we compare the shooting results when the number of shots and number of scores are different?

The shooting results can be represented as a number that allows comparing the number of scores when the number of shots is considered as 1.

Want to explain

④ Let's explain the ways of thinking of the following children.

Nanami's idea

$5 \div 8 = 0.625$
$6 \div 10 = 0.6$

Therefore, we can say Hiroto had the best shooting result.

Daiki's idea

$5 \div 8 = \frac{5}{8}$

$6 \div 10 = \frac{6}{10}$

with common denominator,

$\frac{5}{8} = \frac{25}{40}$ $\frac{6}{10} = \frac{24}{40}$

Therefore, we can say Hiroto had the best shooting result.

**Summary**

If we consider the base quantity (total quantity) as the number of shots and the compared quantity (part of the total) as the number of scores, then we can represent it in the following math sentence.

Shooting result = number of scores ÷ number of shots

$= \frac{\text{number of scores}}{\text{number of shots}}$

Number of shots (Total quantity)
Number of scores (Part of the total)

When comparing the shooting results,

- When the denominators (base quantity) are the same, a result that has a larger numerator (compared quantity) is better.
- When the numerators are the same, a result that has a smaller denominator is better.

Want to try

Let's represent as a number the following results from a basketball game.

① The result when you make 8 scores from 8 shots.

② The result when you make 0 scores from 10 shots.

The number that represents the shooting result becomes a number between 0 and 1.

## 2 The following table shows the number of seats and the number of passengers on airplanes in a given day. Which plane is more crowded?

Number of seats and passengers

| | Number of seats | Number of passengers |
|---|---|---|
| Small plane | 130 | 117 |
| Large plane | 520 | 442 |

The number of seats shows the maximum number of passengers for a safe transport.

The crowdedness is represented as a number that allows comparing the number of passengers when the number of seats is considered as 1.

① Let's find the crowdedness in the small plane.

Compared quantity Base quantity

Number of passengers 0 … 117 130 (people)

Crowdedness 0 … 0.5 … □ 1

| Base quantity | Compared quantity |
|---|---|
| 130 people | 117 people |
| 1 | □ |

Crowdedness

$117 \div 130 =$ □ Crowdedness

② Let's find the crowdedness in the large plane.

Compared quantity Base quantity

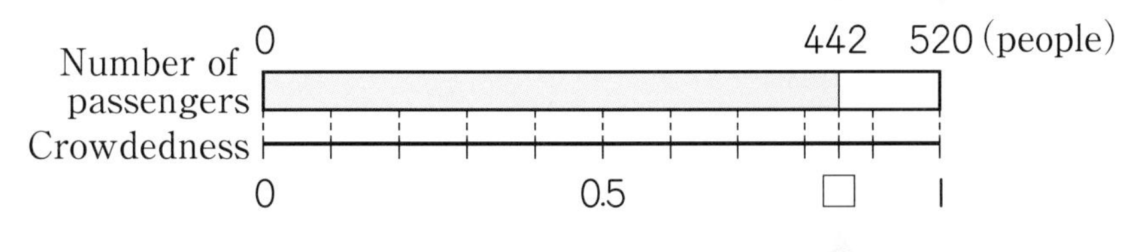


Crowdedness

| Base quantity | Compared quantity |
|---|---|
| 520 people | 442 people |
| 1 | □ |

Crowdedness

□ ÷ □ = □

A number that is expressed by the quantity being compared when the base quantity is considered as 1, like a shooting result or crowdedness, is called a **ratio.**

**Ratio = compared quantity ÷ base quantity**

In particular, it can be represented as shown below when the compared quantity and the base quantity are whole numbers.

$$\text{Ratio} = \frac{\text{compared quantity}}{\text{base quantity}}$$

The ratio of crowdedness of the small plane in the previous page is

$117 \div 130 = 0.9$.

A crowdedness of 0.9 means that the number of passengers is 0.9 when we make the total number of seats as 1.

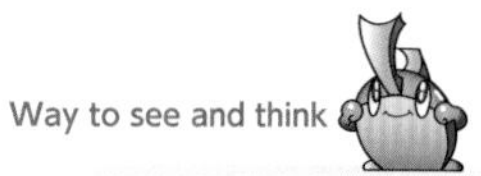


When the base quantity is considered as the unit, the size of the compared quantity is the ratio.

Small plane

| | Base quantity | Compared quantity |
|---|---|---|
| | 130 people | 117 people |
| | 1 | 0.9 (Ratio) |

Large plane

| | Base quantity | Compared quantity |
|---|---|---|
| | 520 people | 442 people |
| | 1 | 0.85 (Ratio) |

**2** Let's find the following ratios.

① The ratio of correct answers when 6 problems out of 10 were answered correctly.

② The ratio of games won when a team won 6 out of 6 soccer games.

③ The ratio of winning lots when someone drew 7 lots which were all blank.

**3** There are 75 children at a party including Manami. Among them, 15 children are from 5th grade. Let's find the ratio of children from 5th grade based on the total number of children at the party.

## 2 Percentage and 歩合 (buai)

Want to know Percentage

**There are 40 passengers in a bus that has 50 seats. Let's think about the crowdedness of this bus.**

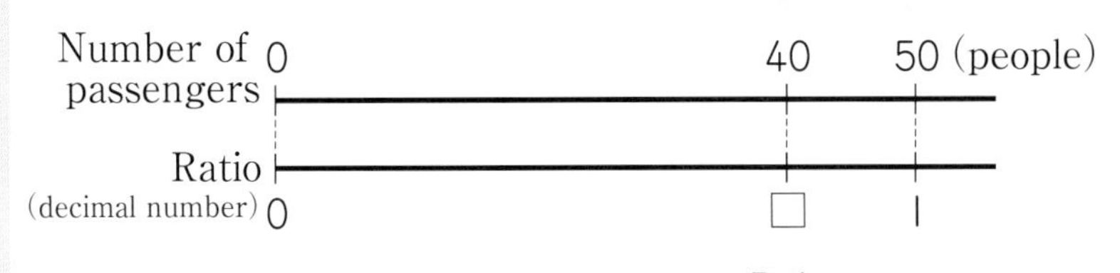


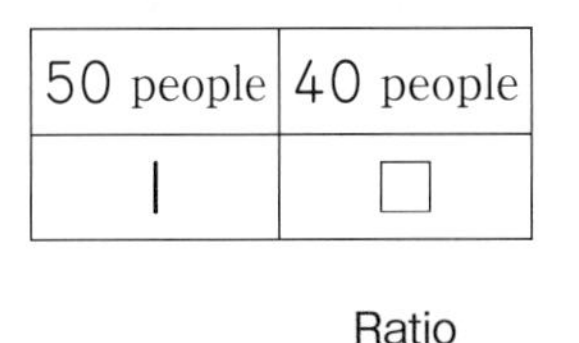

Base quantity Compared quantity

| 50 people | 40 people |
|---|---|
| 1 | □ |

Ratio

① Let's find the ratio of crowdedness of the bus.

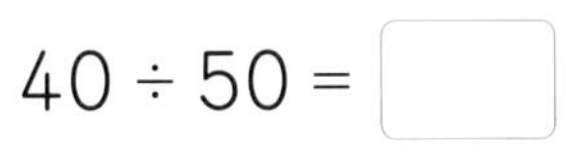

$40 \div 50 = \square$

② Let's represent this ratio by making the base quantity as 100.

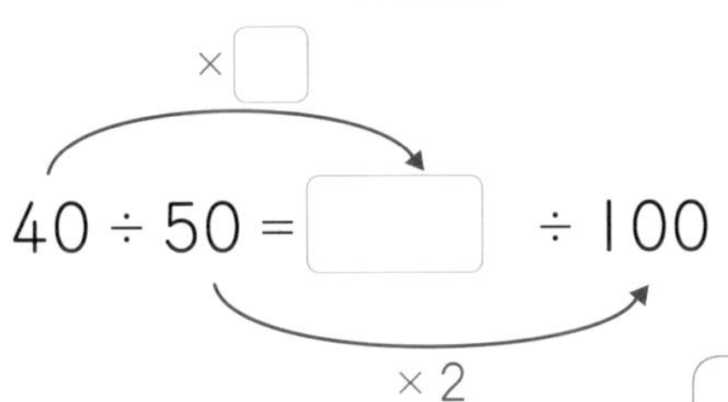


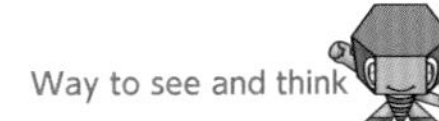


What kind of division rule is used?

We often represent a ratio by making the base quantity as 100. This representation is called **percentage**. The ratio 0.01, which is a decimal number, is called one **percent** and is written as 1 %. If we represent the ratio 1 as a percentage, it is 100 %.

If we multiply a ratio that is represented as a decimal number by 100, it becomes a percentage.

Hiroto

③ Let's represent the crowdedness of the bus as a percentage.

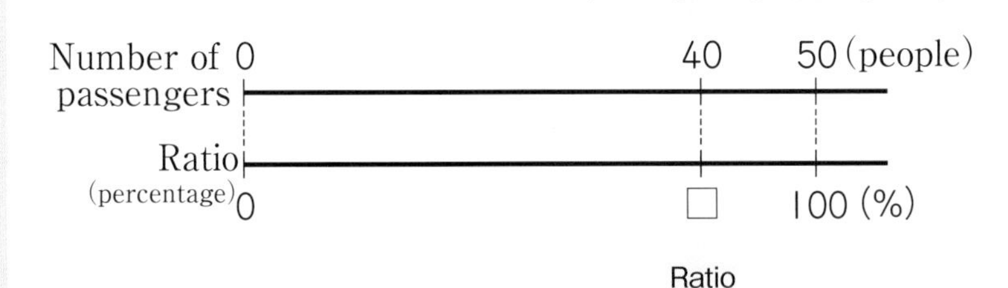


Base quantity Compared quantity

| 50 people | 40 people |
|---|---|
| 1 | □ |
| 100 | □ |

Ratio (percentage)

$40 \div 50 \times 100 = \square$ (%)

Want to confirm

**1** Koki and his friends recorded the vehicles that passed through the road in front of their school for 20 minutes. Let's answer the following questions.

① Let's represent the ratio of each type of vehicle based on the total number of vehicles as a percentage.

② What is the total of all the percentages?

Vehicle survey

| | Number of vehicles | Percentage (%) |
|---|---|---|
| Cars | 63 | 45 |
| Trucks | 35 | |
| Motorcycles | 21 | |
| Buses | 7 | |
| Other | 14 | |
| Total | 140 | |

**2** Let's change the following ratios from decimal numbers to percentages and from percentages to decimal numbers.

① 0.75 ② 0.8 ③ 0.316 ④ 16% ⑤ 2%

Want to try Percentage larger than 100%

**3** There is a train with a capacity of 120 people for each car. The first car has 108 passengers and the second car has 144 passengers. Let's find the crowdedness of each car as a percentage.

① Let's find the crowdedness of the first car.

$108 \div 120 \times 100 =$ ☐ (%)

② Let's find the crowdedness of the second car.

$144 \div 120 \times 100 =$ ☐ (%)

When the number of passengers is more than the capacity, the percentage is larger than 100%.

Want to know 歩合 (buai)

**2** **Yota and his friends held a softball game. The following table summarizes the results of three children. Let's compare the ratio of the number of hits based on the number of bats of the three children.**

Results of softball

| | Number of bats | Number of hits |
|---|---|---|
| Yota | 4 | 1 |
| Airi | 5 | 2 |
| Sho | 5 | 5 |

① Let's find Yota's batting average.

| Number of hits | | Number of bats | | Batting average |
|---|---|---|---|---|
| 1 | ÷ | 4 | = | □ |

② Let's find the batting average of Airi and Sho.

The ratio of the number of hits to the number of bats is called batting average.

In Japanese, the ratio 0.1 is represented as **1 割 (1 wari)**, 0.01 as **1 分 (1 bu)**, and 0.001 as **1 厘 (1 rin)**. These representations are called **歩合 (buai)**.

| 割 | 分 | 厘 |
|---|---|---|
| 0.3 | 5 | 7 |

Yota's batting average of 0.25 is represented in 歩合 (buai) as 2割5分 (2 wari 5 bu).

③ Let's also represent the batting average of Airi and Sho in 歩合 (buai).

Want to confirm

**4** At the supermarket I bought a bento box, that had an original price of 500 yen, for 425 yen. Let's represent, as a percentage and in 歩合 (buai), the ratio of the paid price based on the original price.

What kind of differences are there in the representation of percentages and 歩合 (buai)?

Want to try

**5** In a soccer game, from 5 shots to goal all became scores. Let's represent, as a percentage and in 歩合(buai), the ratio of scores based on the number of shots.

If the ratio written as a percentage is 100%, then that in 歩合(buai) is represented as 10割 (10 wari).

Want to confirm

**6** Let's change the following ratios from decimal numbers to 歩合 (buai) and from 歩合 (buai) to decimal numbers.

① 0.2 ② 0.125 ③ 0.103 ④ 5割2分4厘 (5 wari 2 bu 4 rin)
⑤ 3分7厘 (3 bu 7 rin)

Want to summarize

**7** The following table summarizes the relationship between decimal numbers, percentages, and 歩合 (buai) learned until now. Let's complete the table by filling in the blank spaces.

| Decimal number representing a ratio | 1 | | 0.01 | |
|---|---|---|---|---|
| Percentage | | | | 0.1 % |
| 歩合 (buai) | 10割 (10 wari) | 1割 (1 wari) | | 1厘 (1 rin) |

**That's it**

## Let's find the batting average.

In 2015, a professional baseball player named Shogo Akiyama, set a record for the number of hits in one year. The number of hits and number of bats are shown below. Let's find Akiyama's batting average of that year.

Number of bats: 602 Number of hits: 216

About how much is it in 歩合 (buai)?

## Various fractions

Let's try to reflect on what kind of fractions we have learned until now. Also, let's try to discuss what kind of meaning each has.

① How much of the base size is the size of three parts of a tape that was divided into four equal parts?

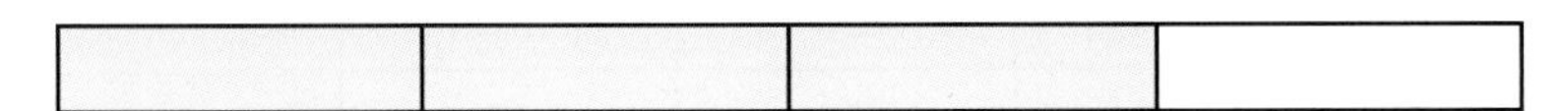

② How many meters is the length of three parts of a 1 m tape that was divided into 4 equal parts?

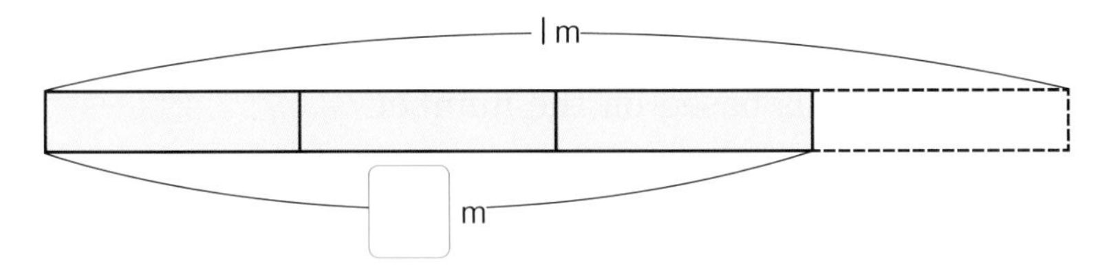


Both ① and ② are three parts of one tape that was divided into four equal parts, but are both the same length?

③ How much is the size of three sets of $\frac{1}{4}$?

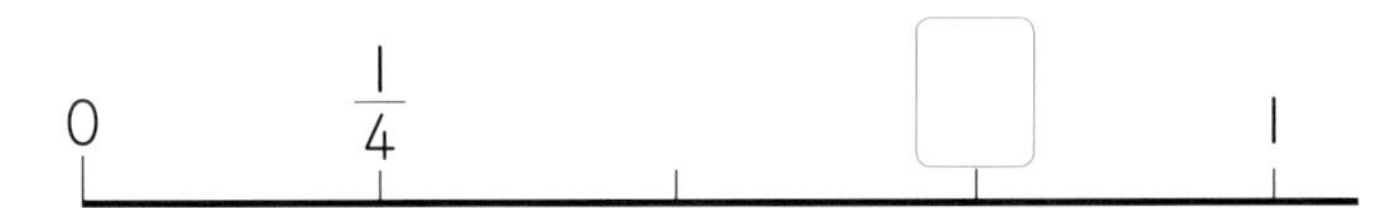


A fraction with a 1 in the numerator, such as $\frac{1}{4}$, is called a unit fraction.

④ From a soccer penalty shotout, let's represent the ratio as a fraction when 3 penalties were scored out of 4 shots.

| Number of shot penalties | Number of scored penalties |
|---|---|
| 4 | 3 |

A ratio can also be represented as a fraction.

⑤ Let's represent the quotient of $3 \div 4$ as a fraction.

$3 \div 4 = \square$

# What you can do now

### Understanding the meaning of a ratio.

**1** In the soccer club, shots were practiced. The table on the right shows the records. Let's answer the following questions with decimal numbers.

| | Number of shots | Number of goals |
|---|---|---|
| Yuki | 15 | 9 |
| Taichi | 12 | 6 |
| Keigo | 20 | 16 |
| Tomoki | 14 | 7 |
| Minoru | 16 | 12 |

① Let's find Yuki's ratio of number of goals based on the number of shots.

② Which students had the same ratio of number of goals based on the number of shots? Also, let's find this ratio.

③ Who had the highest ratio of number of goals based on the number of shots? Also, let's find this ratio.

### Can find a ratio.

**2** Let's find the following ratios.

① The ratio of crowdedness when a bus with a capacity of 60 people has 45 passengers.

② The ratio of shooting goals when all shots are missed out of 15 shots.

③ The ratio of winning lots when someone drew 20 lots and 5 were won.

④ The ratio of correct answers when 10 calculation problems out of 10 were answered correctly.

### Can represent as percentages and in 歩合 (buai).

**3** Let's answer the following questions.

① Let's change the following ratios from decimal numbers to percentages and from percentages to decimal numbers.

0.6　　0.9　　0.105　　65%　　3%　　79%

② Let's change the following ratios from decimal numbers to 歩合 (buai) and from 歩合 (buai) to decimal numbers.

0.7　　0.584　　0.301　　4割3分 (4 wari 3 bu)

1割6分9厘 (1 wari 6 bu 9 rin)

Supplementary Problems p.152

# Usefulness and efficiency of learning

**1** The following table shows Taiga's records for ring toss.

| ○ | ○ | × | × | × | ○ | × | ○ | × | × |
|---|---|---|---|---|---|---|---|---|---|

○ scored × missed

Let's answer the following questions when Taiga's results is represented as 0.4.

① What does the number 0.4 represent?

② Let's represent Taiga's result as a percentage.

③ If he tosses two more rings and scores both, what will his result be?

④ When is the result of ring toss represented as 0?

☐ Understanding the meaning of a ratio.

☐ Can find a ratio.

**2** Let's answer the following ratios as percentages and in 歩合 (buai).

① In basketball practice, the ratio of shooting success when 20 shots were scored out of 25 shots.

② The ratio of germination when 30 buds sprout out of 30 flower seeds.

③ The ratio of crowdedness when a bus with a capacity of 50 people has 26 passengers.

④ The ratio of correct answers when 7 problems out of 10 were answered correctly.

☐ Can represent as percentages and in 歩合 (buai).

**3** There is a train with a capacity of 80 people for every car. Let's answer the following questions.

① Let's find the crowdedness of the first car as a percentage when it carries 60 passengers.

② Let's find the crowdedness of the second car as a percentage when it carries 96 passengers.

☐ Can find a ratio.

**4** In a fireworks festival, 18000 shots are fired. Among those, 6300 shots are "star mine" fireworks. Let's find, as a percentage, the ratio of "star mine" fireworks based on the total fireworks.

☐ Can find a ratio.

Chikugo River Fireworks Display
(Kurume City, Fukuoka Prefecture)

# The area is the same?

Problem How should we find the area of the inside of the frame?

# 14 Area of Figures

## Let's think about how to find the area.

### 1 Area of parallelograms

Want to know: How to find the area of parallelograms

Activity

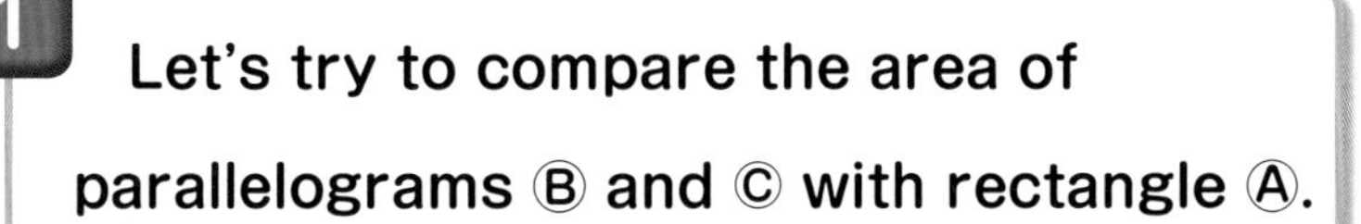

1
Let's try to compare the area of parallelograms Ⓑ and Ⓒ with rectangle Ⓐ.


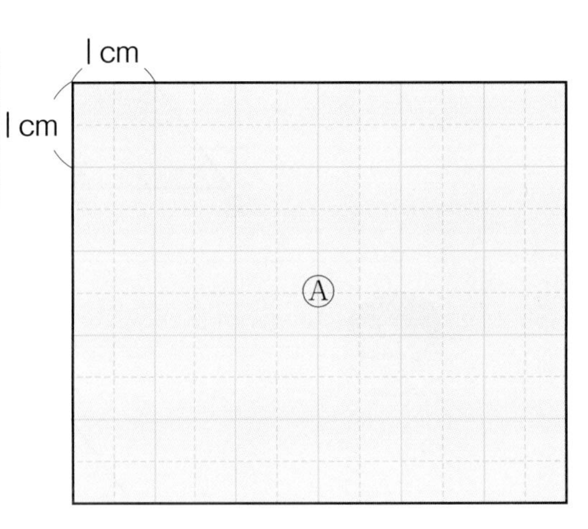


① Let's try to measure the length of each side of the quadrilaterals Ⓐ, Ⓑ, and Ⓒ.

② Let's try to compare the area of the quadrilaterals Ⓐ, Ⓑ, and Ⓒ.

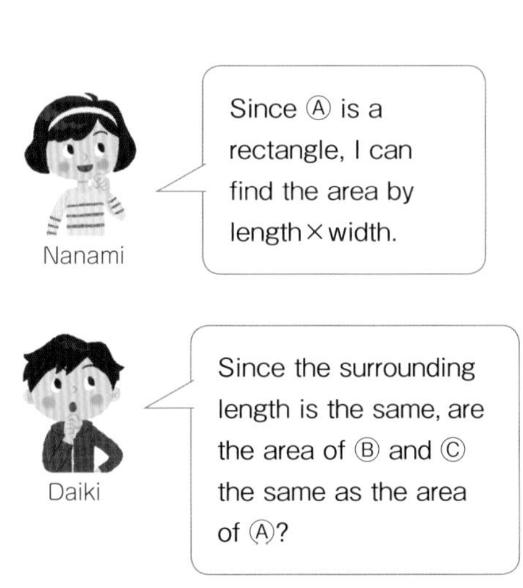


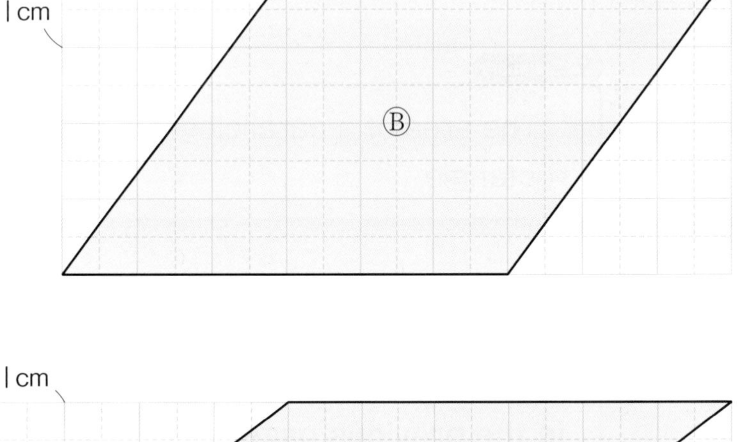

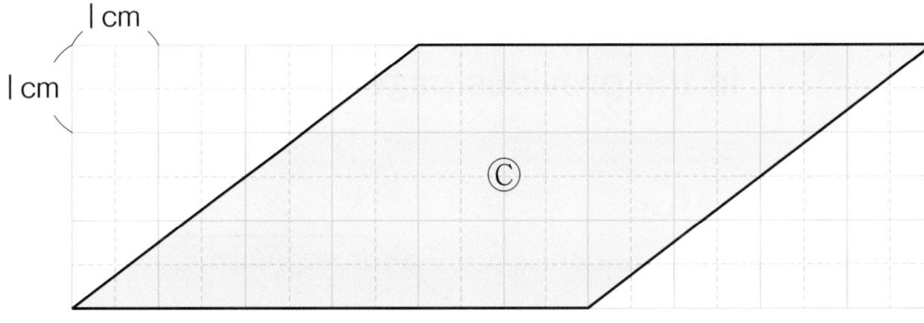


**Purpose** How should we find the area of a parallelogram?

Want to explain

③ Nanami and Hiroto changed the shape of the parallelogram Ⓑ as shown below. Let's explain the ideas of the following children.

Nanami's idea

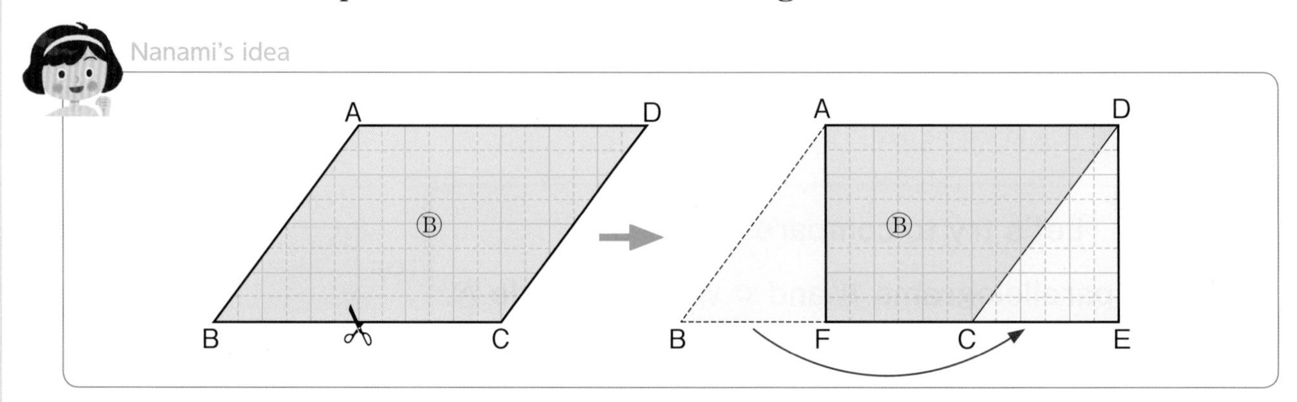


Hiroto's idea

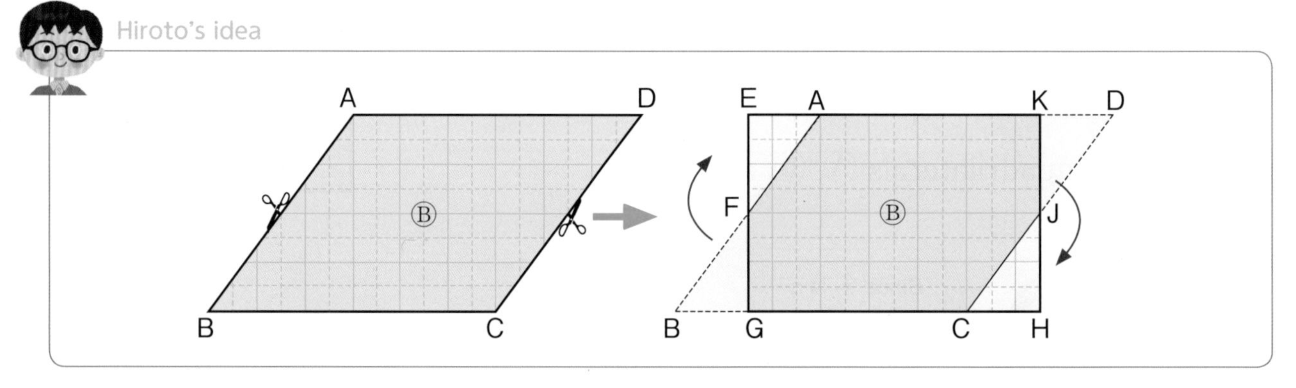


Summary

The area of a parallelogram can be found by changing the figure into a rectangle.

Want to think

**2** **Let's think about how to calculate the area of parallelogram Ⓒ shown in the previous page.**

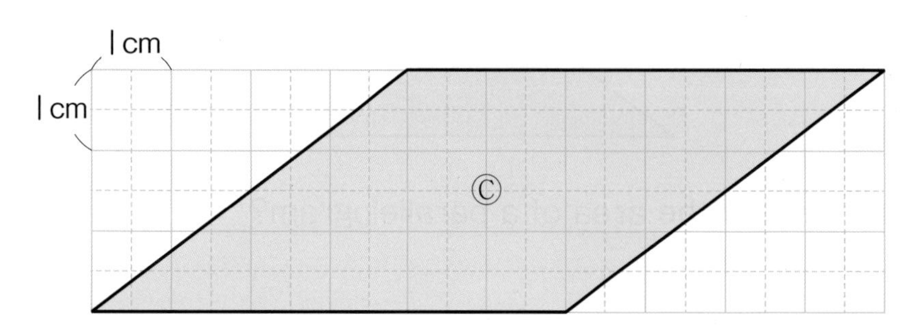


The area of a rectangle can be found with length × width and the area of a square with side × side.

Purpose Is there a formula to find the area of a parallelogram?

① Which lengths need to be identified to find the area of Ⓒ?

Let's draw straight lines inside the diagram.

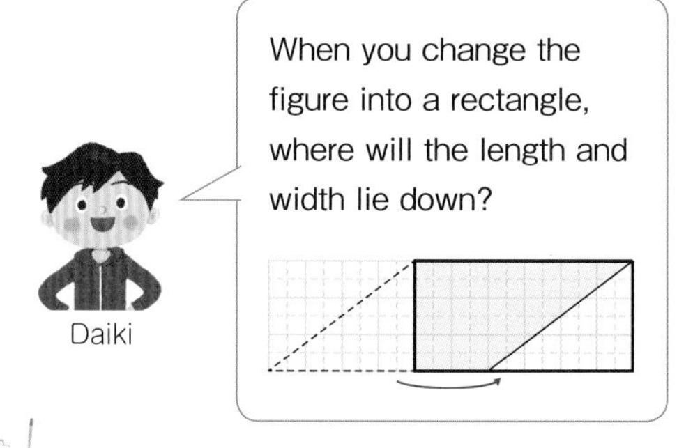

When you change the figure into a rectangle, where will the length and width lie down?
Daiki


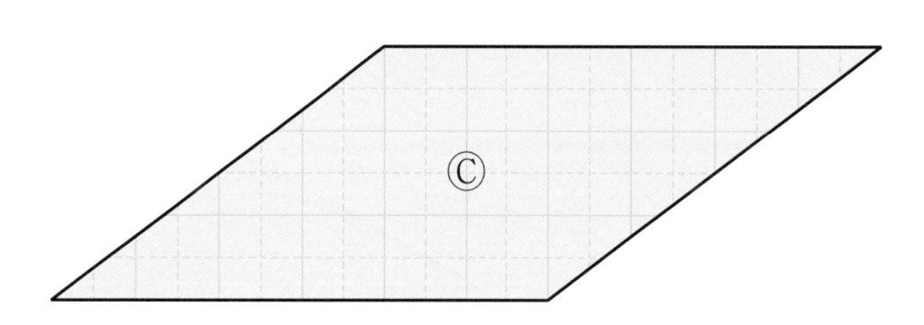


Think of side BC as the **base** of the parallelogram shown on the right. When we draw straight line AG, straight line EF, and other straight lines, which are perpendicular to base BC, the length of these straight lines is called the **height** against the base BC.

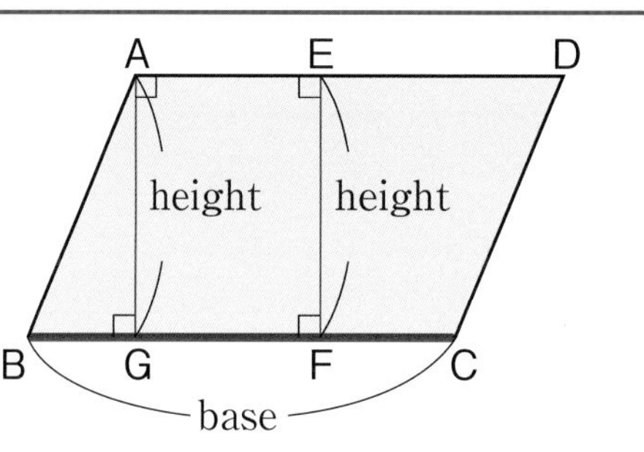


Summary

If you know the base and height, the area of a parallelogram can be found with the following formula.

**Area of a parallelogram = base × height**

Want to confirm

**1** Let's find the area of Ⓐ, Ⓑ, and Ⓒ in page 43 by using the formula. Also, let's compare each of the sizes.

**2** Let's find the area of the following parallelograms.

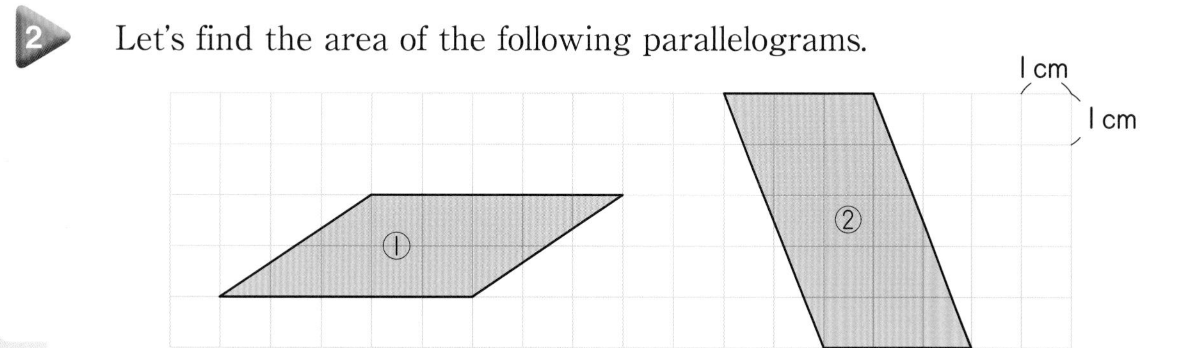

## 3 Let's find the area of the following parallelogram.

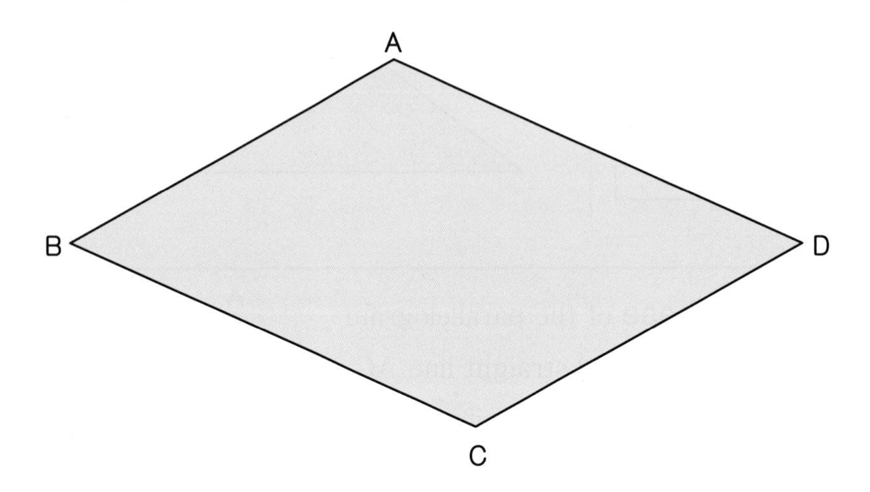


① When side BC is the base, let's find the area by measuring the height.

Area = ☐ × ☐ = ☐ ($cm^2$)

② When side CD is the base, let's find the area by measuring the height.

Area = ☐ × ☐ = ☐ ($cm^2$)

The height depends on the base.

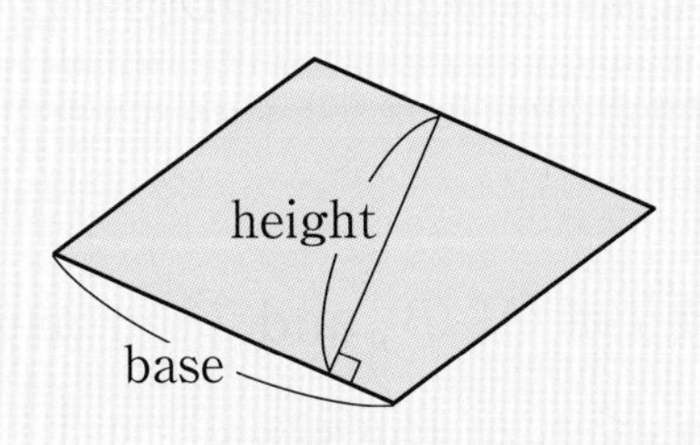


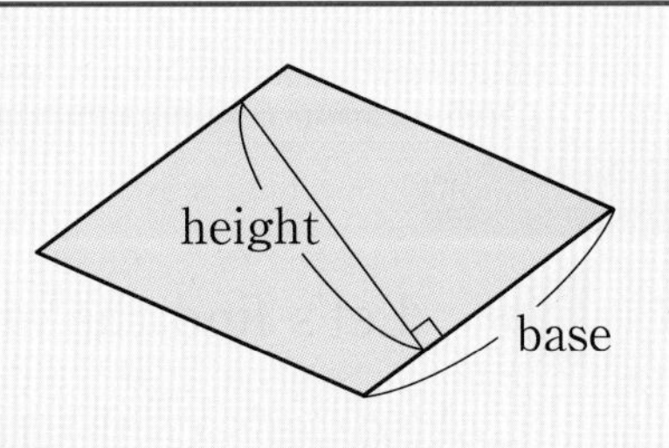


Want to confirm

**3** Let's find the area of the following parallelograms.

①

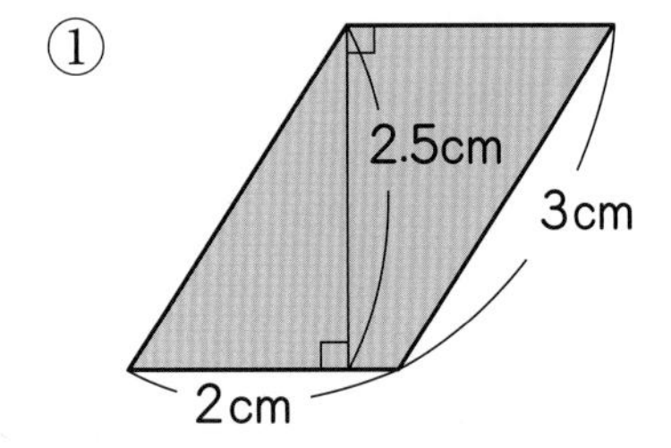


②

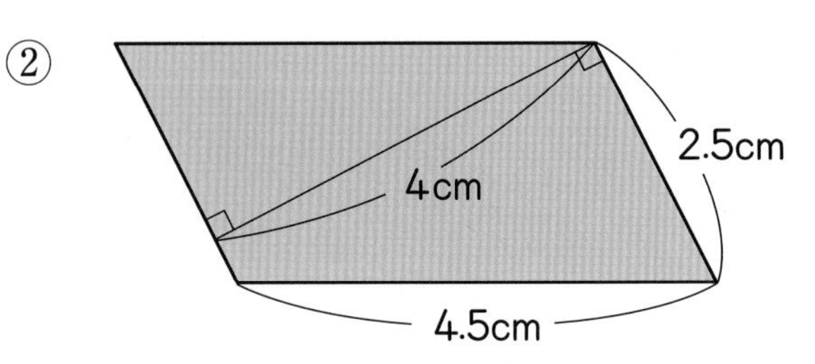

Want to know The place of the height

Activity

## 4

**Let's think about how to find the area of the parallelogram shown on the right when side BC is the base.**

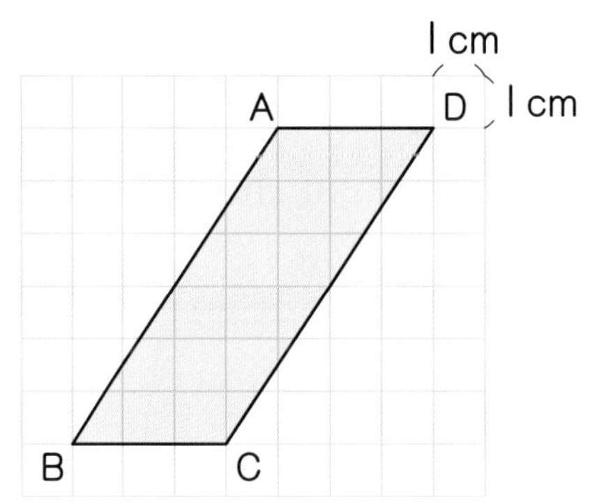

**Purpose** Where is the height of a parallelogram?

Want to explain

① Let's explain the ideas of the following children.

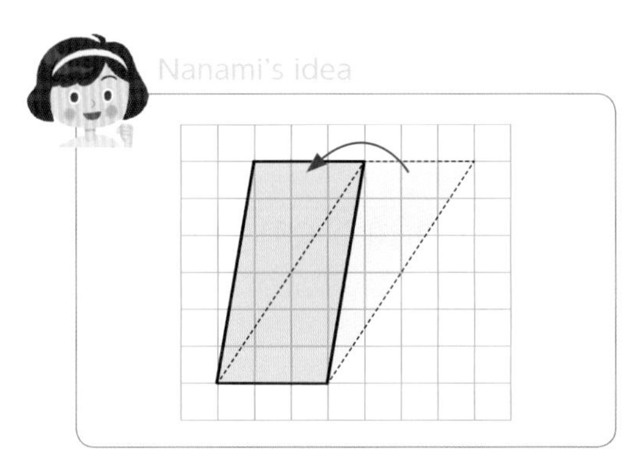


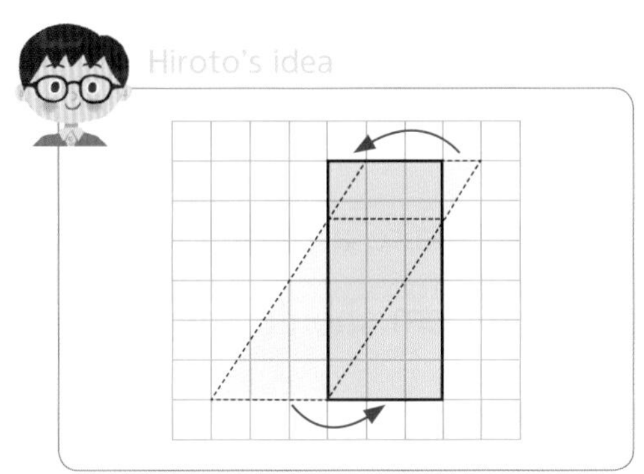


② Let's find the area of parallelogram ABCD.

**Summary**

When side BC is the base, the distance between straight line Ⓐ and straight line Ⓑ is the height of parallelogram ABCD.

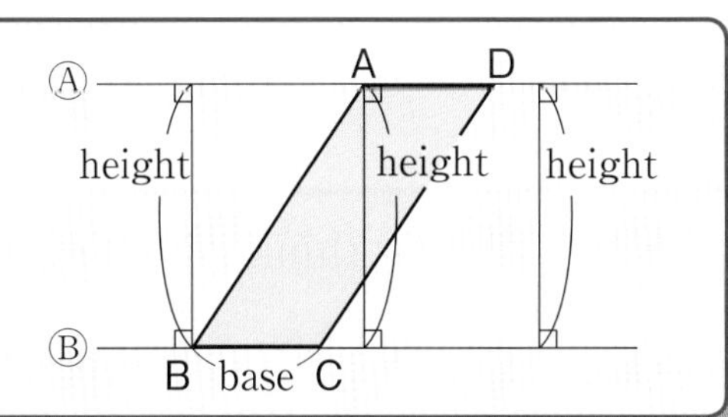


Want to confirm

Let's find the area of the following parallelograms.

①

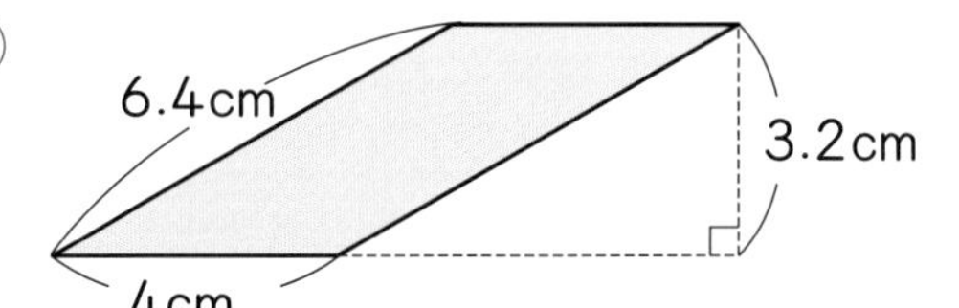


②

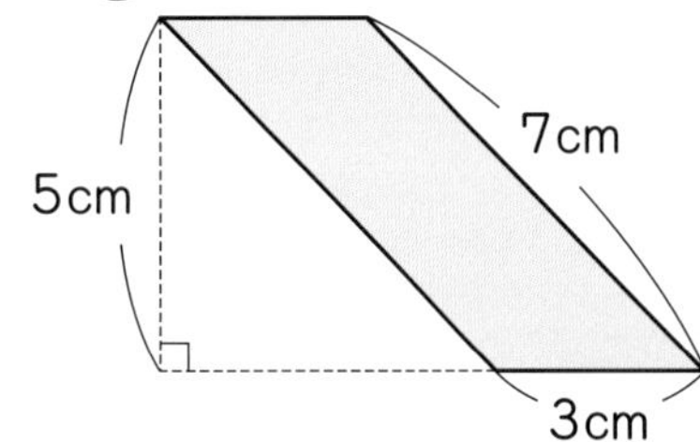

Want to explore

**5** Let's find the area of parallelograms Ⓐ, Ⓑ, and Ⓒ. Also, let's compare each of the areas and explain the understood things.

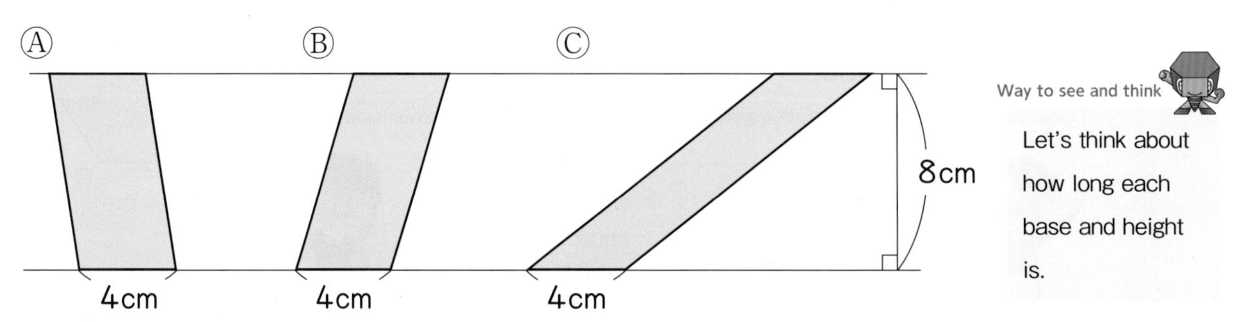


Way to see and think

Let's think about how long each base and height is.

In any parallelogram, if the lengths of the bases are equal and the heights are also equal, the areas become equal.

Want to try

**6** We will make a parallelogram with an area of 48 $cm^2$ and a height of 8 cm. How many cm should the length of the base be?

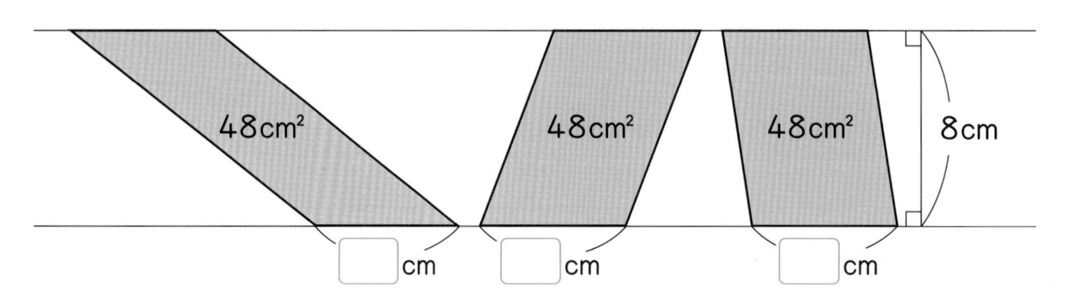


Let's think by using the formula to find the area of a parallelogram.

☐ × 8 = 48

Base Height Area

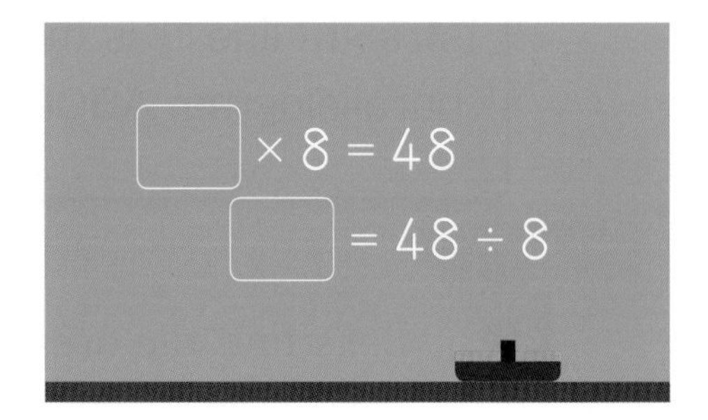

## 2 Area of a triangle

Want to know How to find the area of a triangle

Activity

**1 Let's think about how to find the area of the following triangle.**

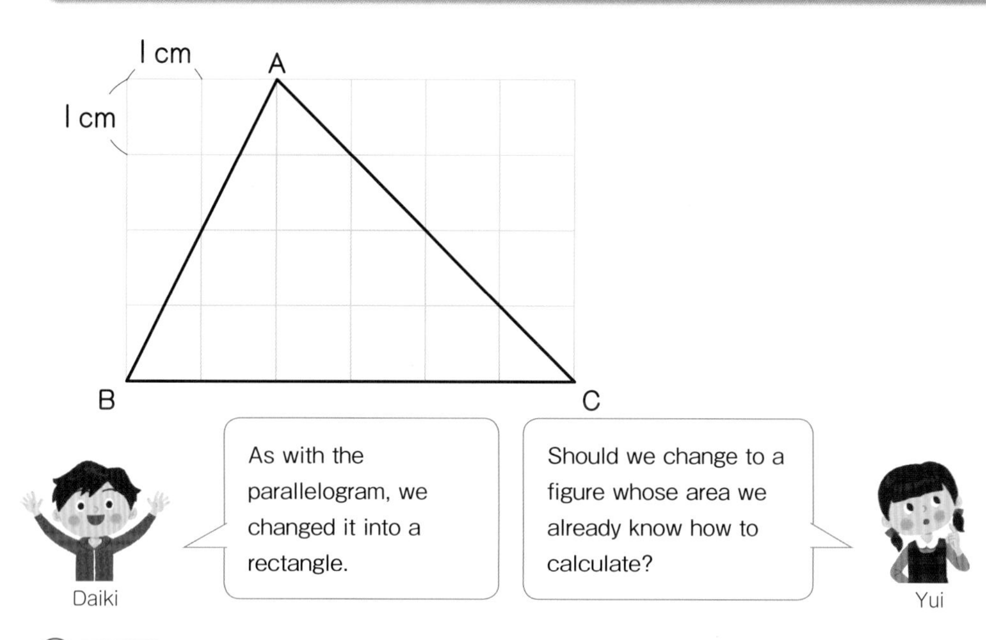

1 cm
1 cm
A
B
C
As with the parallelogram, we changed it into a rectangle.
Daiki
Should we change to a figure whose area we already know how to calculate?
Yui

Purpose How should we find the area of a triangle?

Want to think

① Let's think by drawing in the diagram.

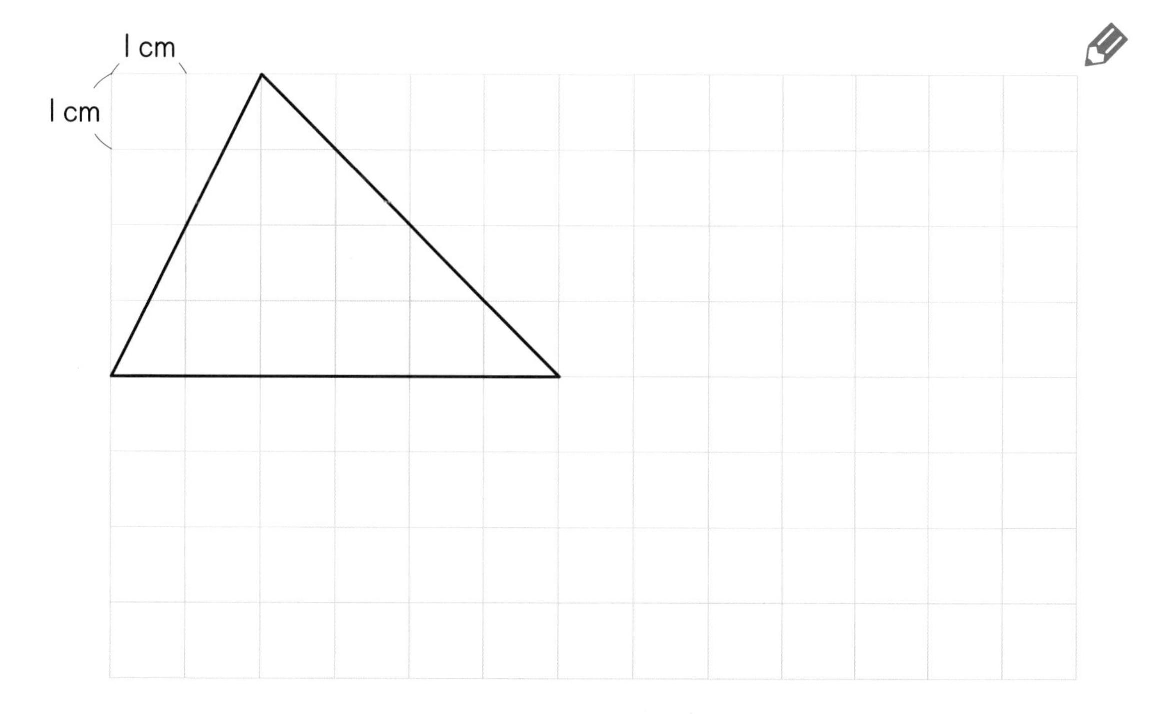

Want to explain

Activity

② Let's explain the ideas of the following children.

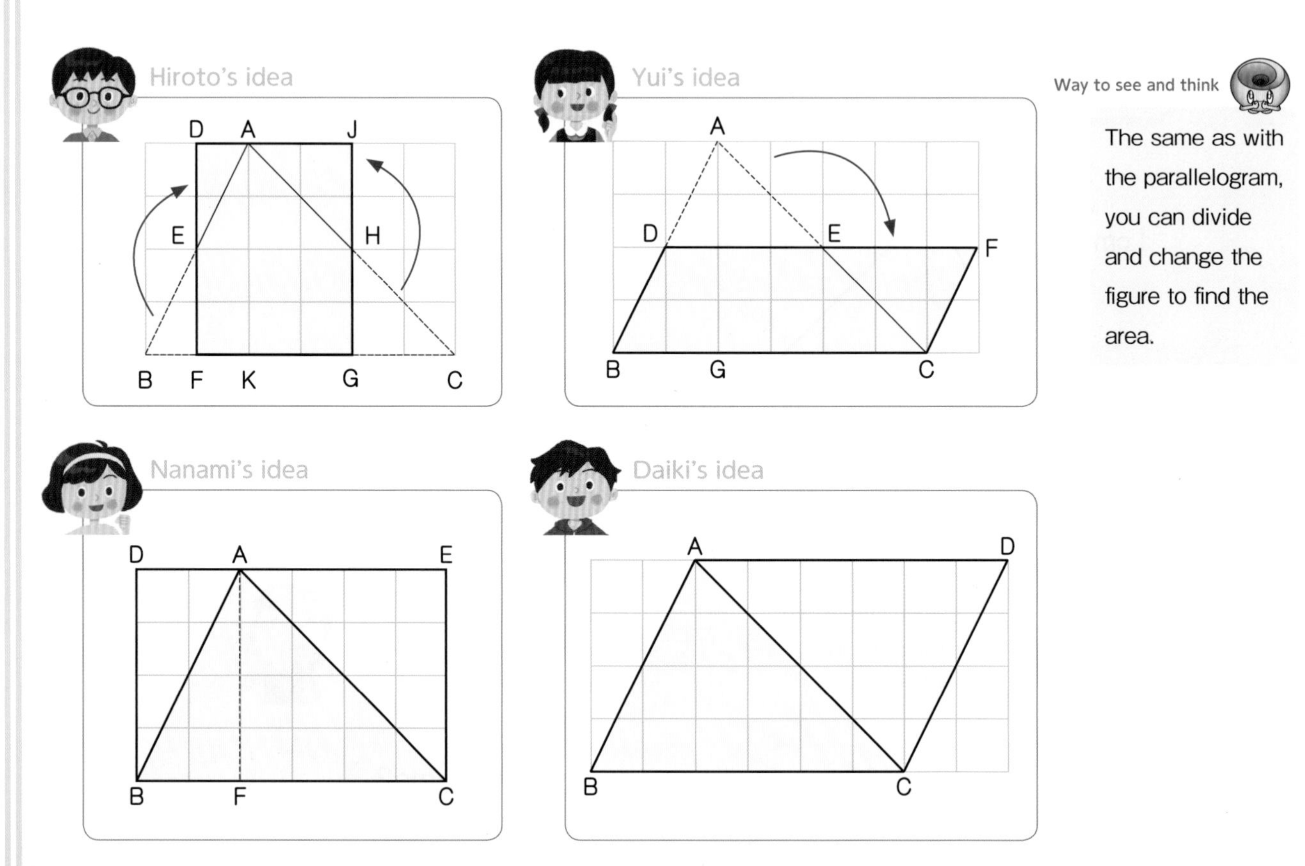


Way to see and think

The same as with the parallelogram, you can divide and change the figure to find the area.

Want to discuss

③ Among the ideas of the children in ②, let's discuss similarities and differences.

I considered to change it into a rectangle.
It was changed to a figure with an equal area.
I changed it into a parallelogram.
I changed it to a figure with 2 times of the area.

Summary

If you change a triangle into a rectangle or a parallelogram, you can find its area.

Activity

## 2 Let's think about the formula to find the area of triangle 1 in page 49.

① Let's think based on the ideas of the children from ② in the previous page.

**Hiroto's idea**

Since the width of the rectangle is half of BC,

AK × (BC ÷ 2)

4 × (6 ÷ 2)

**Yui's idea**

Since the height of the parallelogram is half of AG,

base × (AG ÷ 2)

6 × (4 ÷ 2)

**Way to see and think**

Let's try to summarize each of the ideas, and find what they have in common.

**Nanami's idea**

Since the area of the triangle is half of the area of rectangle DBCE and the length of the rectangle is AF,

(AF × BC) ÷ 2

(4 × 6) ÷ 2

**Daiki's idea**

Since the area is half of the area of parallelogram ABCD,

(base × height) ÷ 2

(6 × 4) ÷ 2

② Let's discuss similarities and differences about the math expressions considered in ①.

**Purpose** Is there a formula to find the area of a triangle?

③ Considering the four diagrams from ② in the previous page, where is the same length as the original triangle?

Think ob side BC as the **base** of the triangle shown on the right. When we draw straight line AD, which is perpendicular to side BC, from vertex A that is opposite to side BC, the length of straight line AD is called the **height** against the base BC.

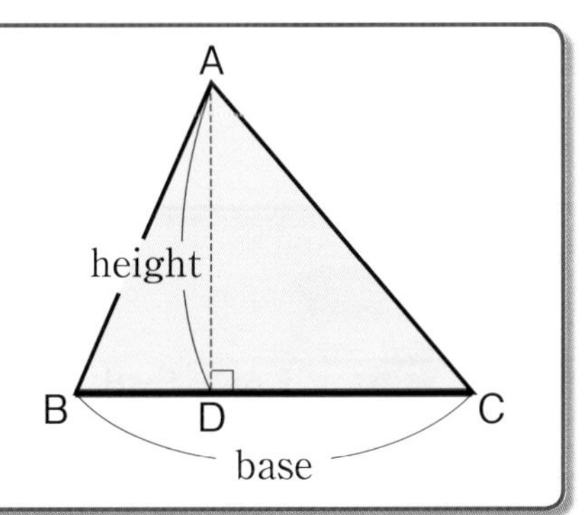


**Summary**

If you know the base and height, the area of a triangle can be found with the following formula.

**Area of a triangle = base × height ÷ 2**

**1** Let's find the area of the following triangles.

①

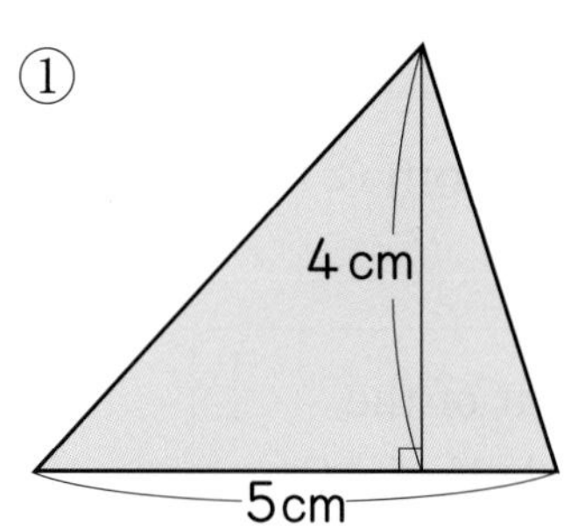


②

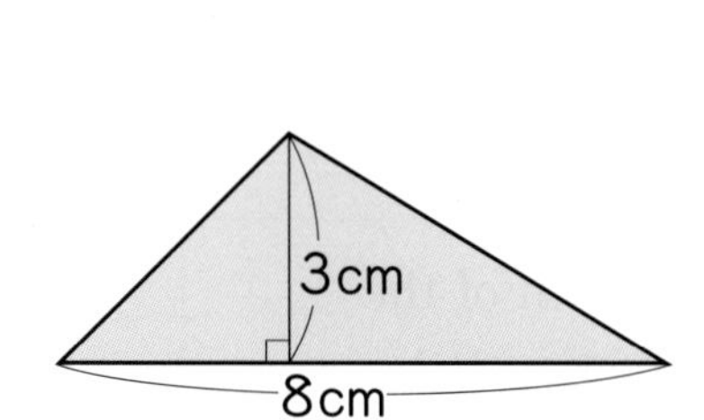

**2** Let's measure the necessary lengths and find the area of the following triangle.

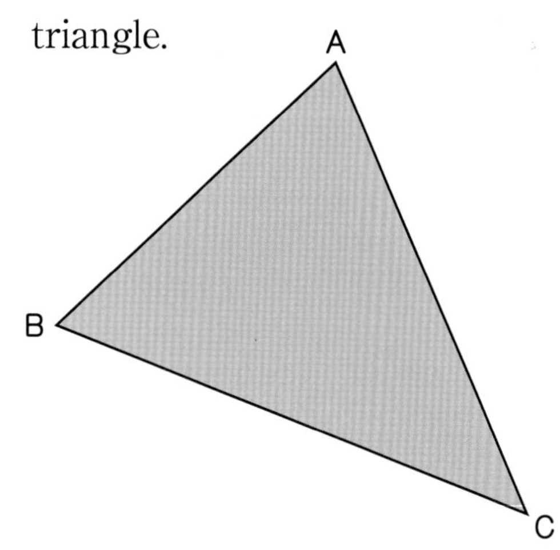

The same as with the parallelogram, you can consider various sides as the base.

The height depends on the base.

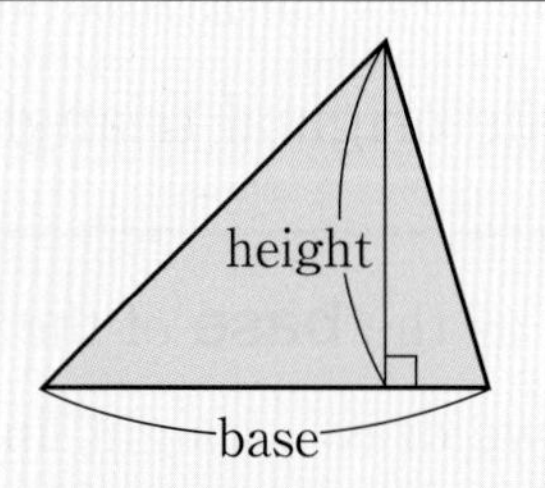


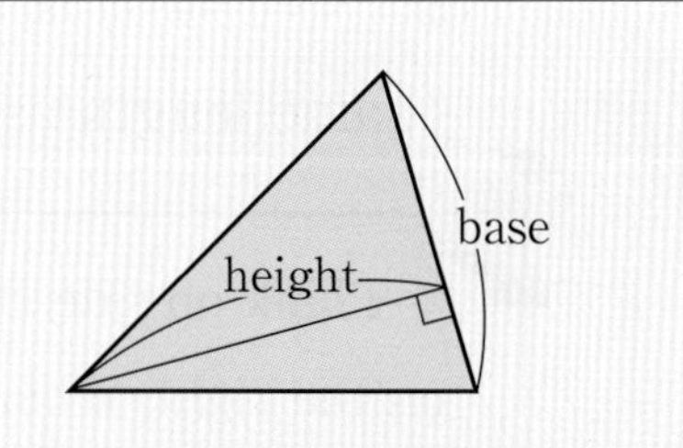

**3** Let's find the area of triangle ABC shown on the right considering the following.

① When side BC is the base.

② When side AB is the base.

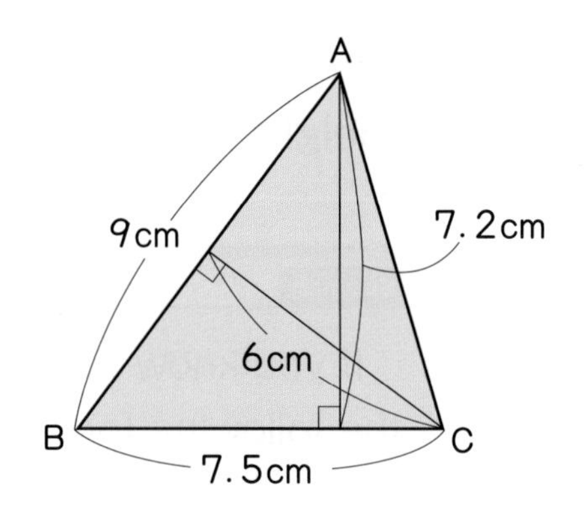

Want to know The place of the height

Activity

## 3 Let's think about how to find the area of the triangle with side BC as the base shown on the right.

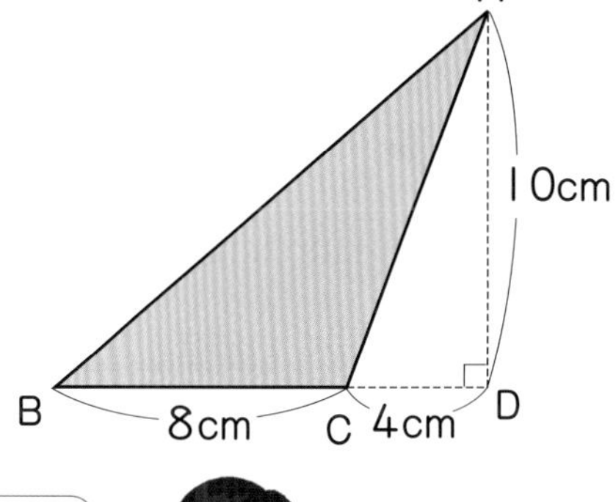


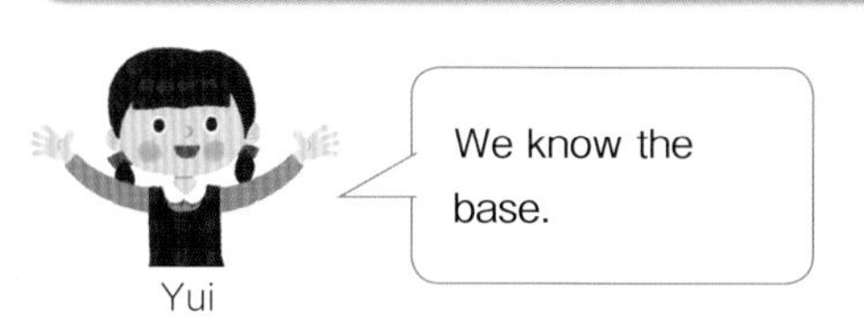


Where was the height when it was a parallelogram?

**Purpose** Where is the height of a triangle?

Want to explain

① Let's explain the ideas of the following children and find the area with each of the ideas.

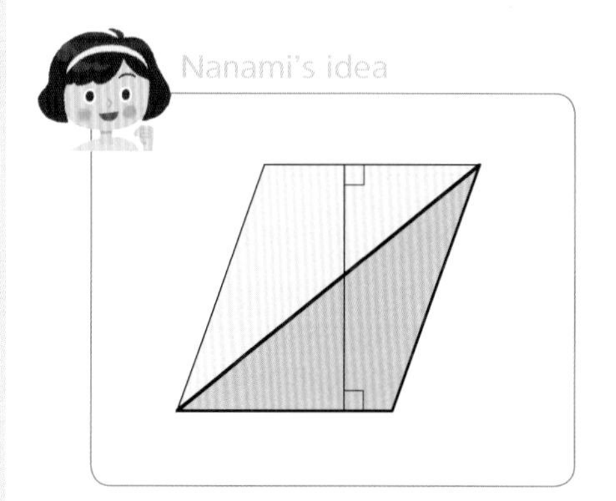


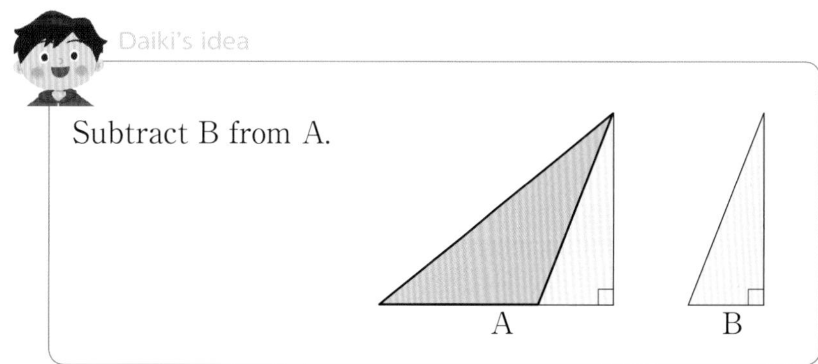


② Let's find the area of the triangle that has a base of 8 cm and a height of 10 cm by using formulas and compare it with the answer in ①.

**Summary**

Draw a straight line Ⓐ through vertex A and parallel to side BC, and a straight line Ⓑ tracing side BC. The distance between straight lines Ⓐ and Ⓑ is the height of the triangle when side BC is the base.

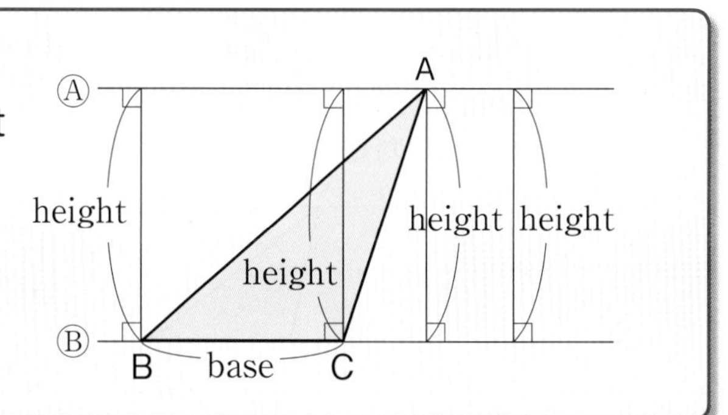

Want to confirm

**4** Let's find the area of the triangles shown on the right.

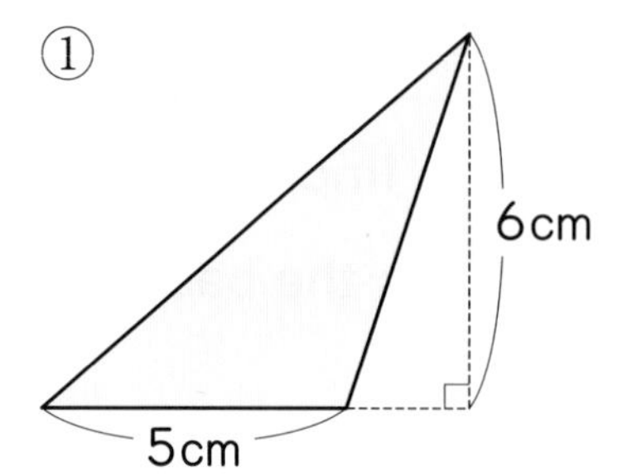


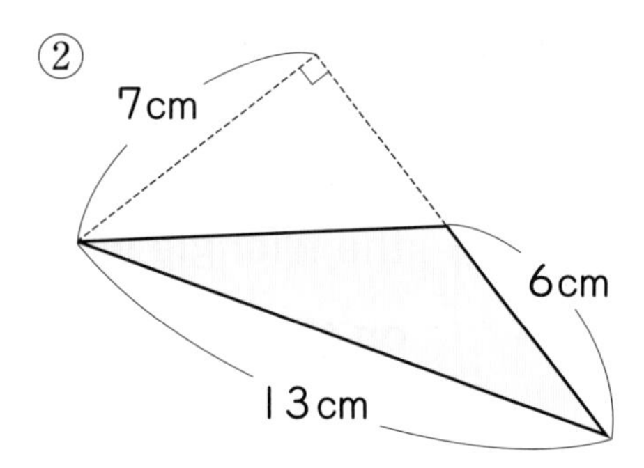


Want to explain

**5** In the following diagram, straight lines Ⓔ and Ⓕ are parallel. Let's find the area of each triangle. Also, let's compare each of the areas and explain the understood things.

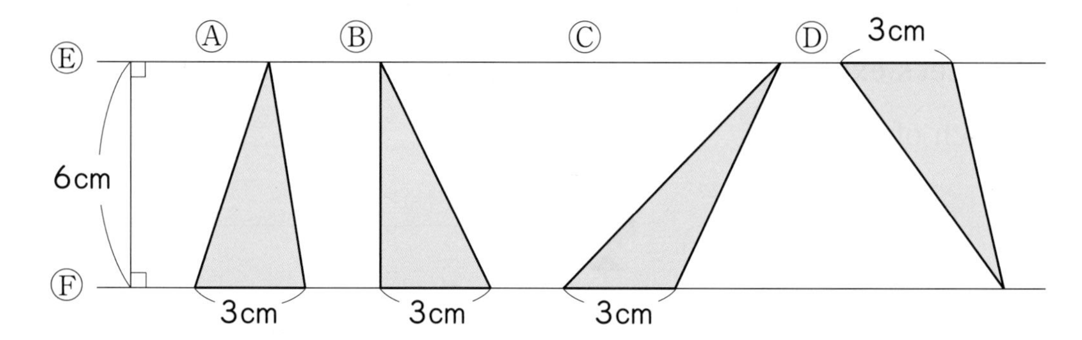

In any triangle, if the lengths of the bases are equal and the heights are also equal, the areas become equal.

Want to represent

**6** In the following diagram, straight lines Ⓓ and Ⓔ are parallel. Let's explain the reasons why the triangles Ⓐ, Ⓑ, and Ⓒ have equal area by using the words "base" and "height."

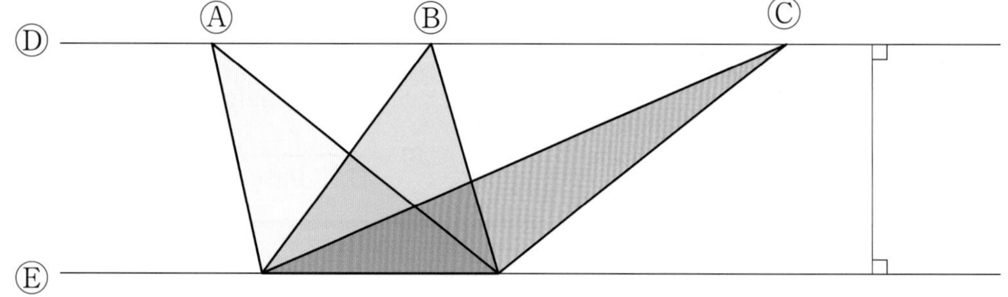


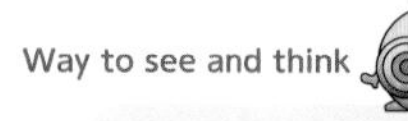


Let's explain clearly and in order the reasons why the areas are equal.

Want to think

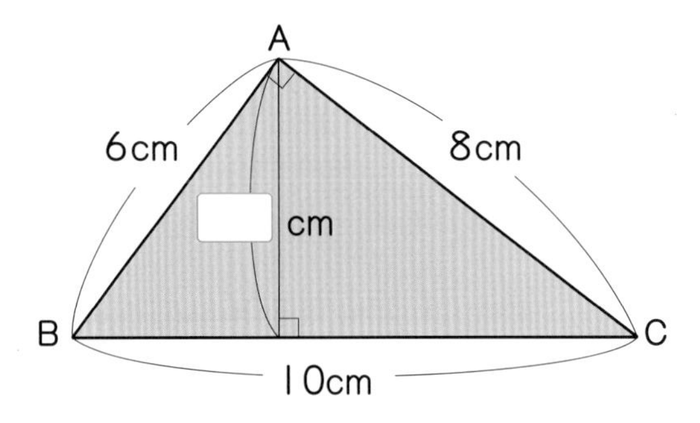


## 4 Let's think about the right triangle shown on the right.

① Let's find the area.

② Let's calculate the height of the triangle when side BC is the base.

10 × ☐ ÷ 2 = Area

Base    Height

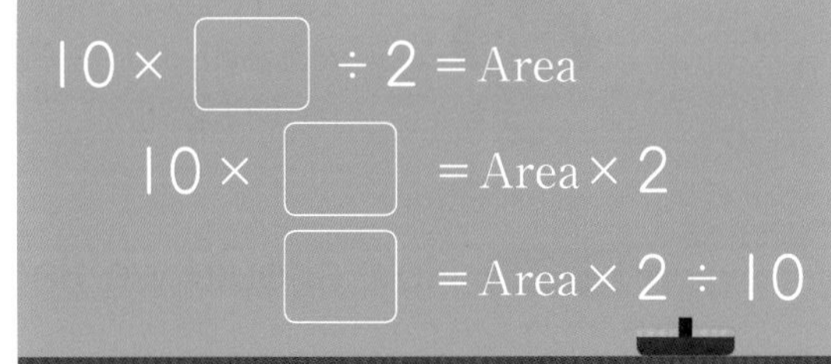


Want to confirm

7 The triangle shown on the right has an area of 84 cm².

How many cm is the height when the base is 14 cm?

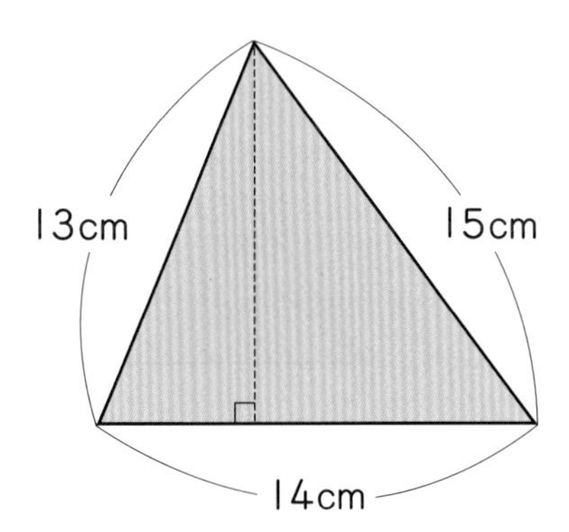


Want to deepen

8 Let's find the height of each triangle when sides AD and BC are the respective bases.

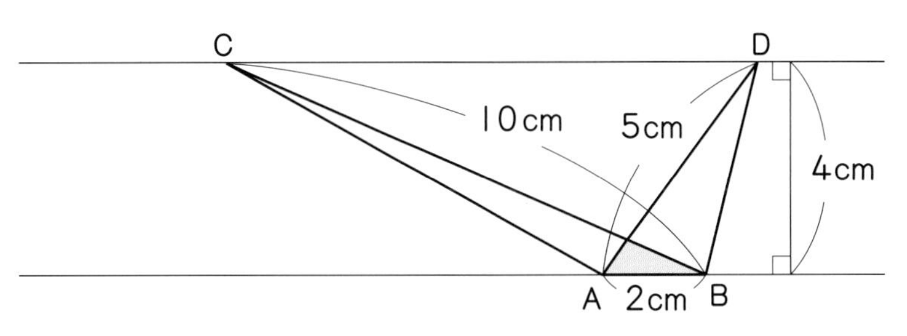

## 3 Area of a trapezoid

Want to know — How to find the area of a trapezoid

Activity

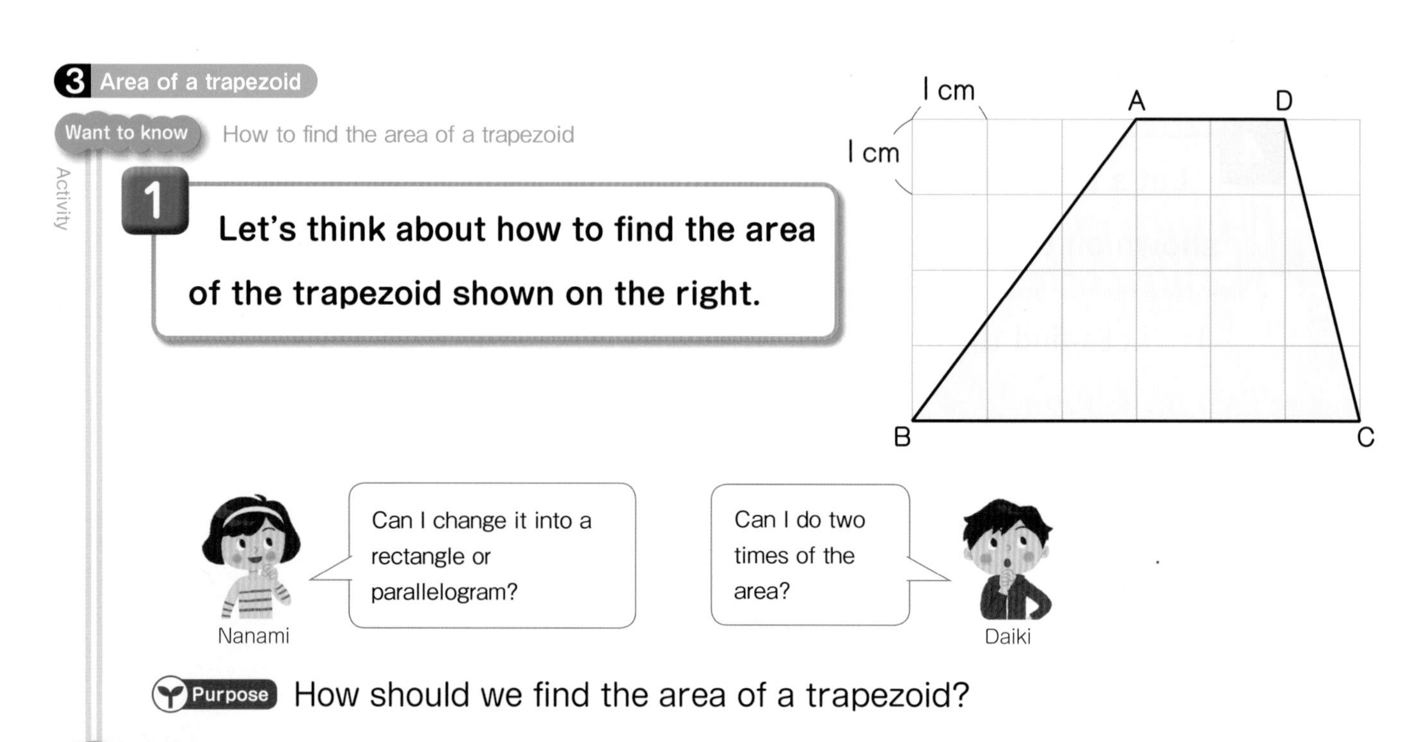


**1** **Let's think about how to find the area of the trapezoid shown on the right.**

Purpose How should we find the area of a trapezoid?

Want to explain

① Let's explain the ideas of the following children.

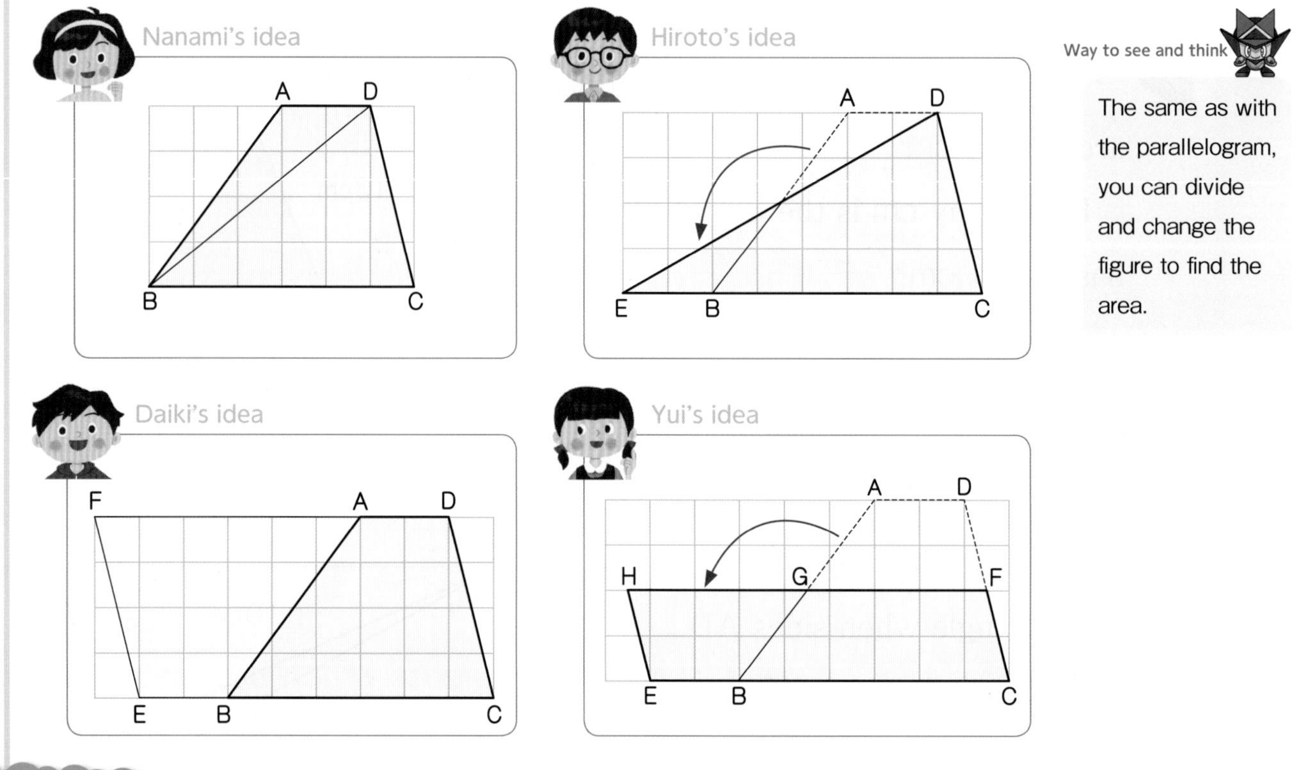


Way to see and think

The same as with the parallelogram, you can divide and change the figure to find the area.

Want to compare

② Among the ideas of the children in ①, what are the similarities and differences?

Want to think

③ From the ideas in ①, let's think about the formula to find the area of a trapezoid.

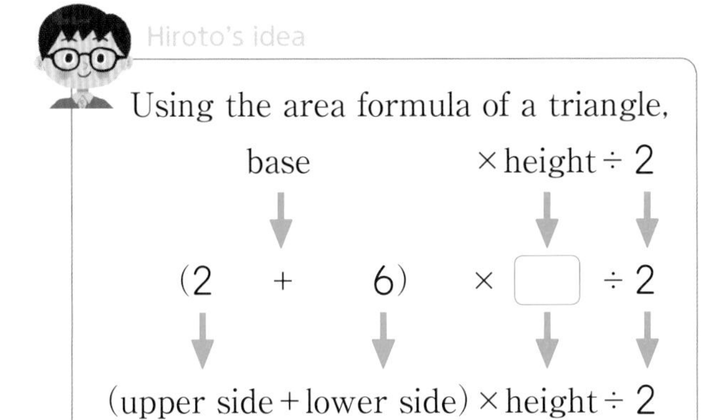

Hiroto's idea
Using the area formula of a triangle,
base ×height ÷ 2
(2 + 6) × □ ÷ 2
(upper side + lower side) × height ÷ 2


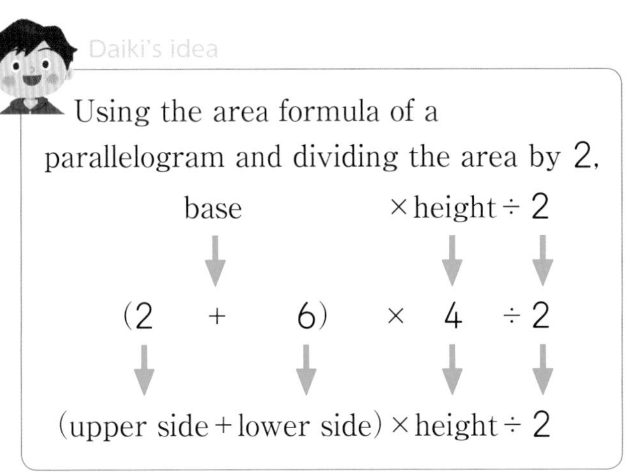

Daiki's idea
Using the area formula of a parallelogram and dividing the area by 2,
base ×height ÷ 2
(2 + 6) × 4 ÷ 2
(upper side + lower side) × height ÷ 2


The two parallel sides of the trapezoid are called **upper base** and **lower base**, and the distance between them is called the **height**.

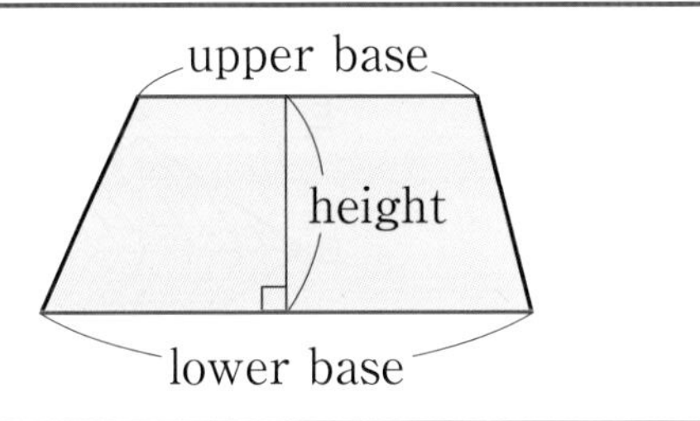


Summary

If you know the upper base, lower base and height, the area of a trapezoid can be found with the following formula.

**Area of a trapezoid = (upper base + lower base) × height ÷ 2**

Want to confirm

1 Let's find the area of the following trapezoids.

①

1 cm
1 cm

②

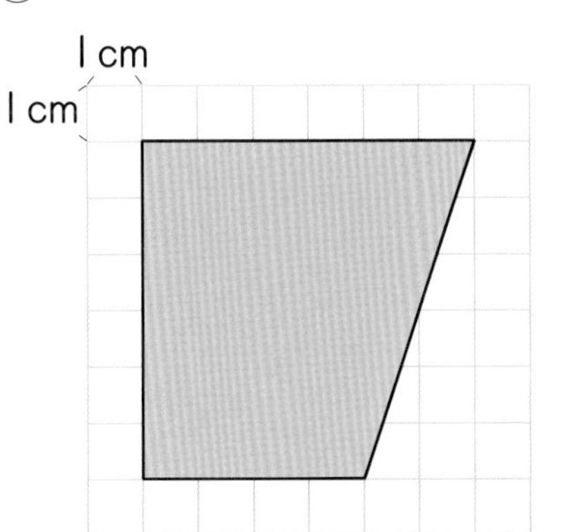


③

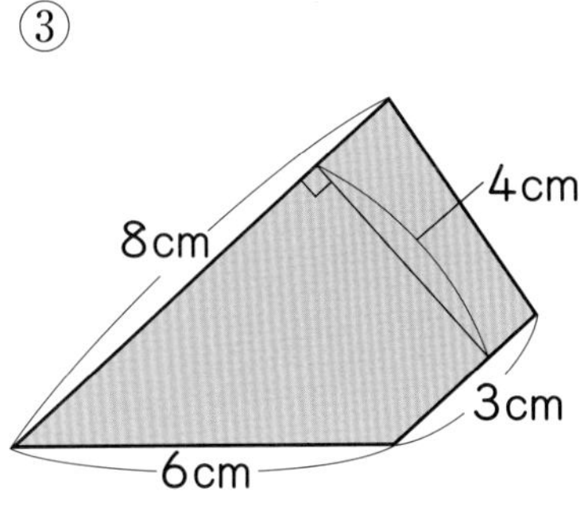

## 4 Area of a rhombus

Want to know How to find the area of a rhombus

Activity

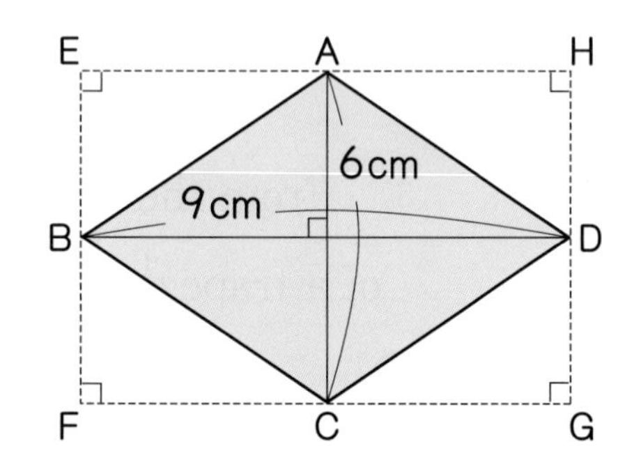


**1** **Let's think about how to find the area of the rhombus shown on the right.**

Just change it to another figure.

Can I calculate it?

Purpose How can we find the area of a rhombus?

Want to explain

① Let's explain the ideas of the following children.

Nanami's idea

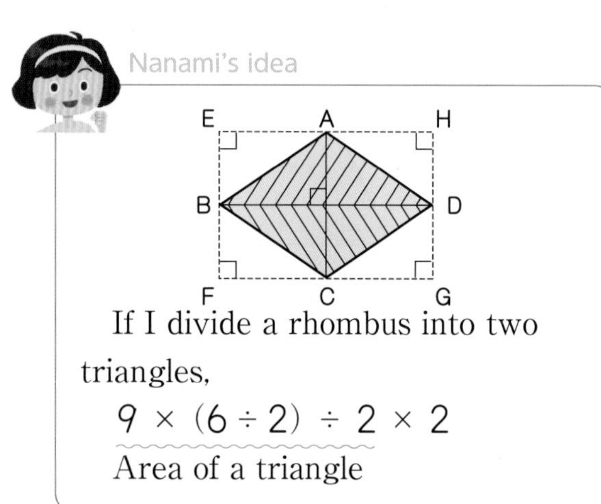


If I divide a rhombus into two triangles,

$9 \times (6 \div 2) \div 2 \times 2$

Area of a triangle

Hiroto's idea

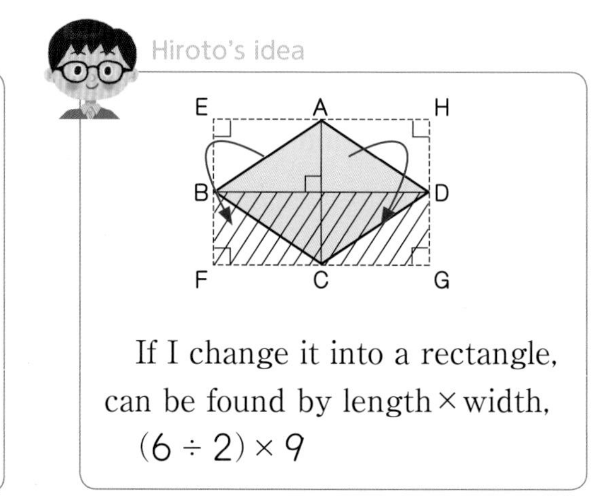


If I change it into a rectangle, can be found by length × width,

$(6 \div 2) \times 9$

Daiki's idea

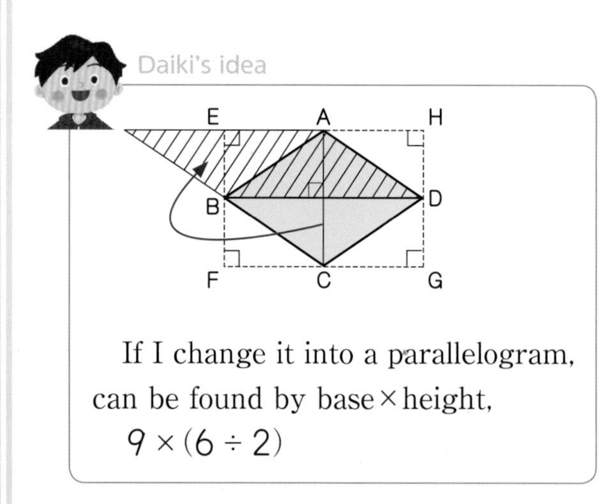


If I change it into a parallelogram, can be found by base × height,

$9 \times (6 \div 2)$

Yui's idea

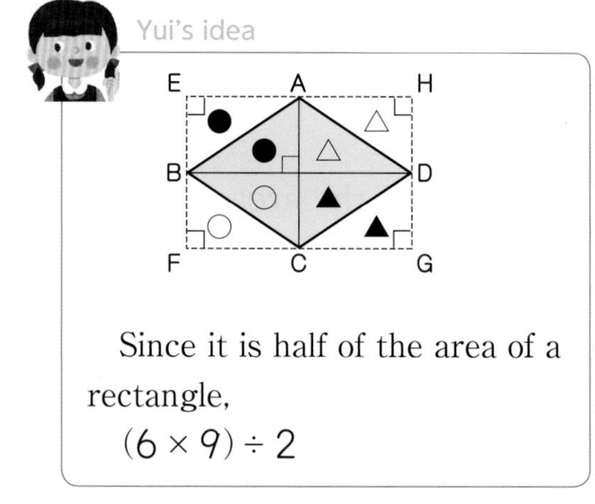


Since it is half of the area of a rectangle,

$(6 \times 9) \div 2$

② From the ideas in ①, let's think about the formula to find the area of a rhombus.

**Summary**

If you know the length of the two diagonals, the area of a rhombus can be found with the following formula.

**Area of a rhombus = diagonal × diagonal ÷ 2**

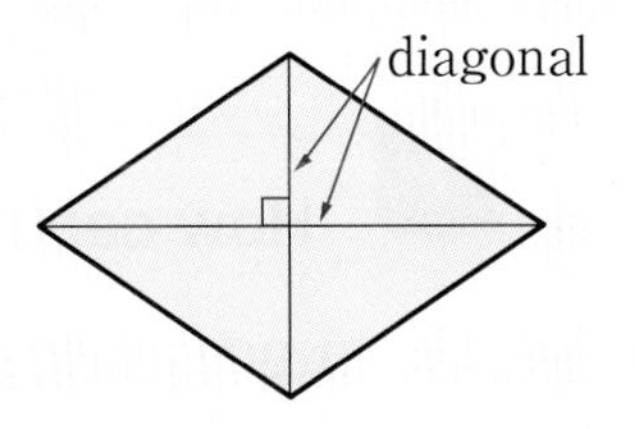


Want to deepen

Hiroto considered, as shown in the diagram on the right, the area of the rhombus from 1 in the previous page. Let's explain Hiroto's way of thinking.

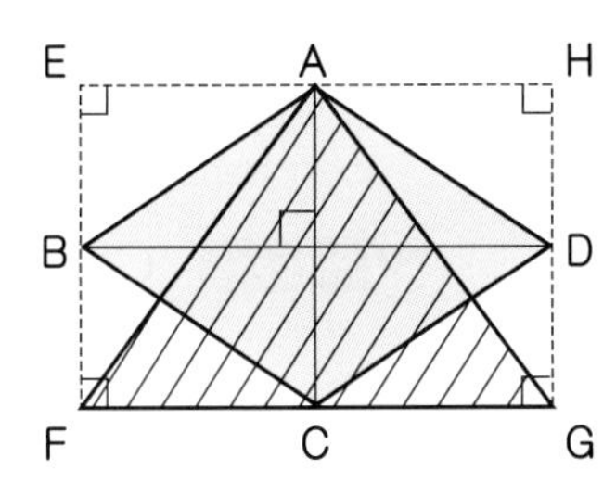


Want to confirm

Let's find the area of the following rhombuses.

①

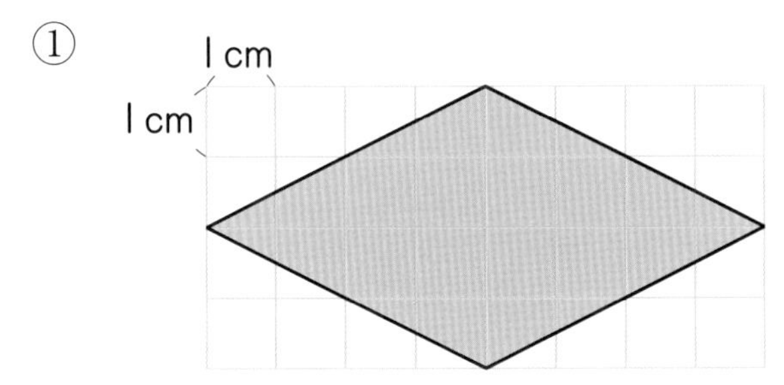


②

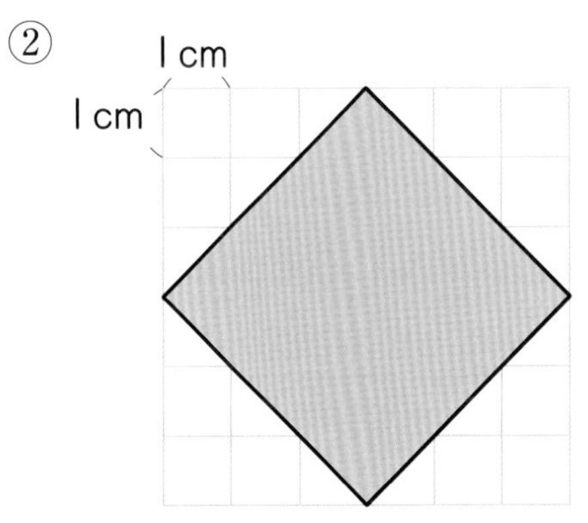


③

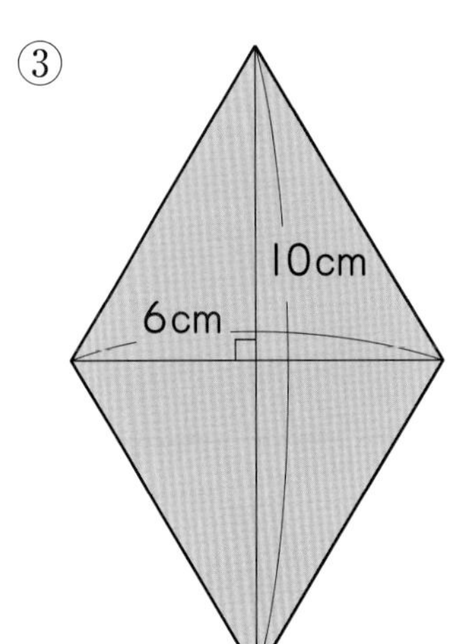


④

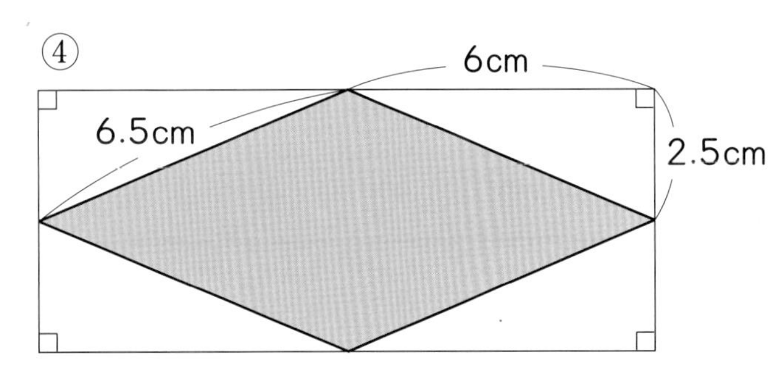


Want to try

Let's think about how to find the area of a quadrilateral with diagonals that have a perpendicular intersection as shown on the right.

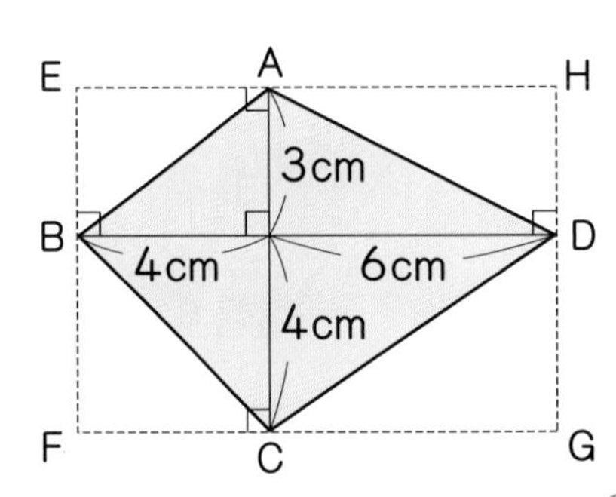

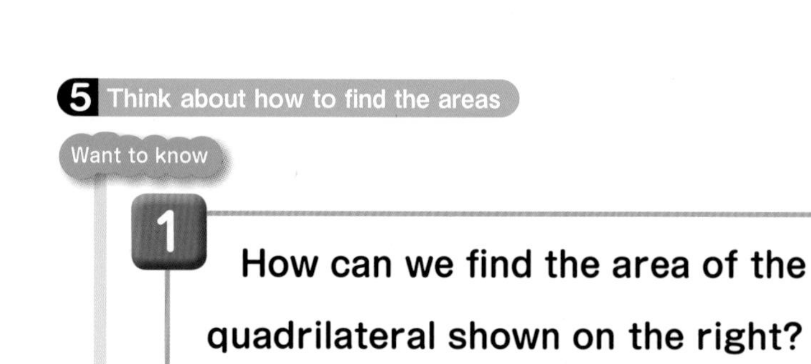

5 Think about how to find the areas
Want to know
1
How can we find the area of the quadrilateral shown on the right?

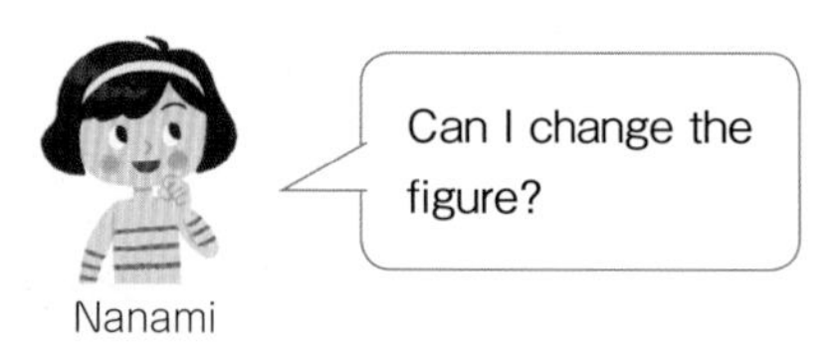

Can I change the figure?
Nanami


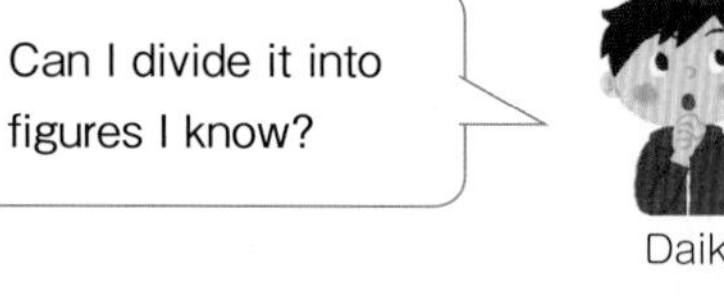

Can I divide it into figures I know?
Daiki


**Purpose** Can we also find the area of various quadrilaterals?

Want to explain

① Yui divided the figure as shown on the right and found the area. Let's explain Yui's idea.

② Let's measure the necessary lengths and find the area.

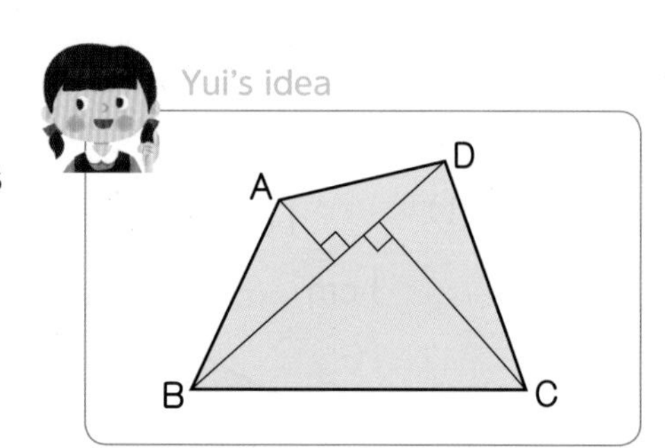


**Summary**

The area of a quadrilateral or a pentagon can be found by dividing them into several triangles.

Want to try

**1** Let's find the area of the following pentagons.

①

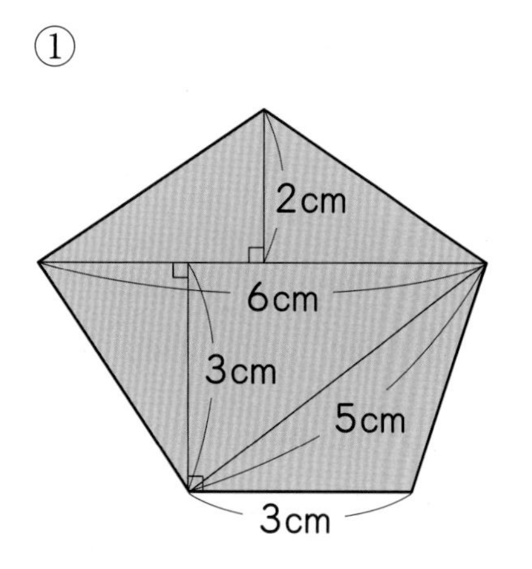


②

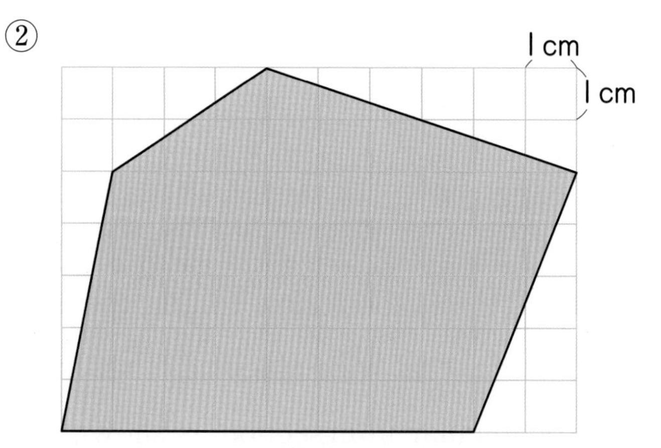

Want to deepen

Let's think about various ways to find the colored area shown on the right.

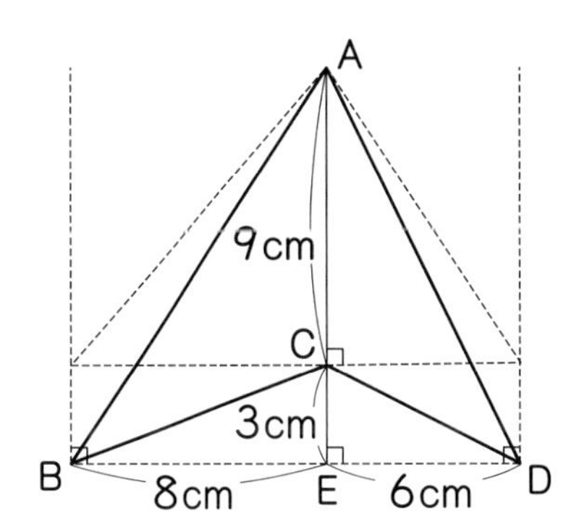


Want to confirm

Let's find the area of the following figures.

①

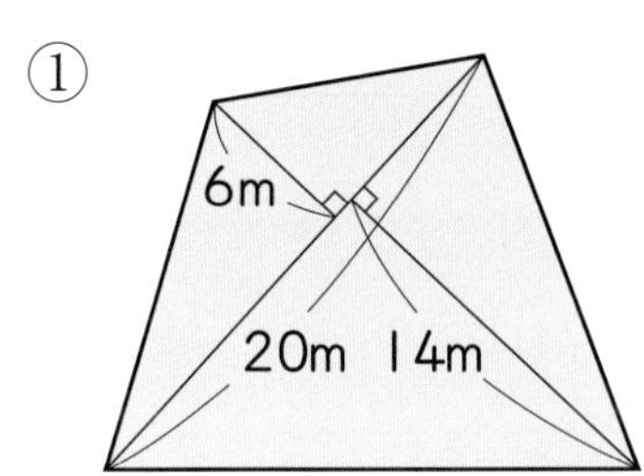


②

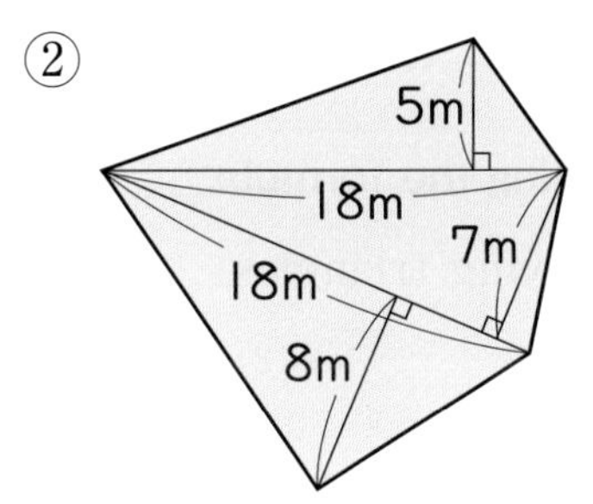


③

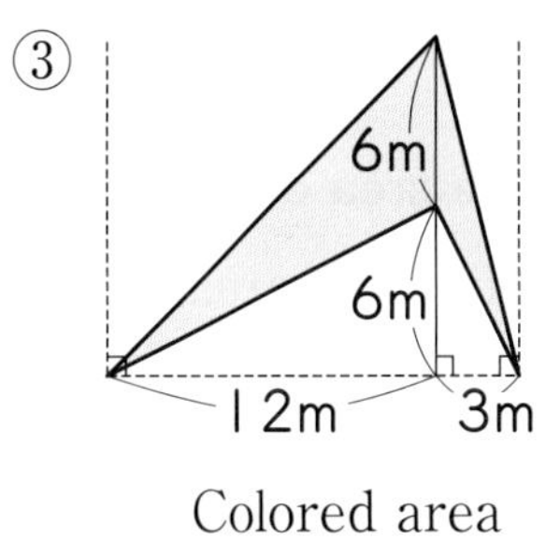


Colored area

Want to deepen

As shown in the diagram on the right, the height of the triangle is increased by 1 cm. Let's answer the following questions.

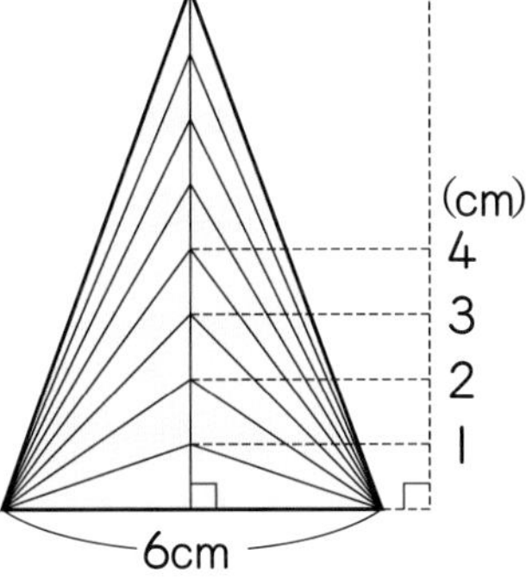


① Let's write the area formula of a triangle and explore the quantities changing together.

② Let's summarize the relationship between the height and the area in the following table.

Height and area of the triangle

| Height (cm) | 1 | 2 | 3 | | | | | | | |
|---|---|---|---|---|---|---|---|---|---|---|
| Area ($cm^2$) | 3 | | | | | | | | | |

③ Can you say that the area is proportional to the height? Let's also write the reasons.

④ Let's write a math sentence considering the height as □ cm and the area as ○ $cm^2$.

⑤ How many cm is the height when the area of the triangle is 30 $cm^2$?

# What you can do now

☐ **Can find the area of parallelograms by using the formula.**

**1** Let's find the area of the following parallelograms.

① 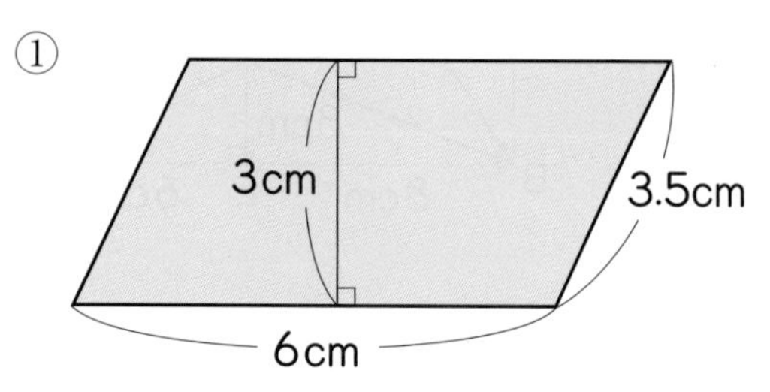


② 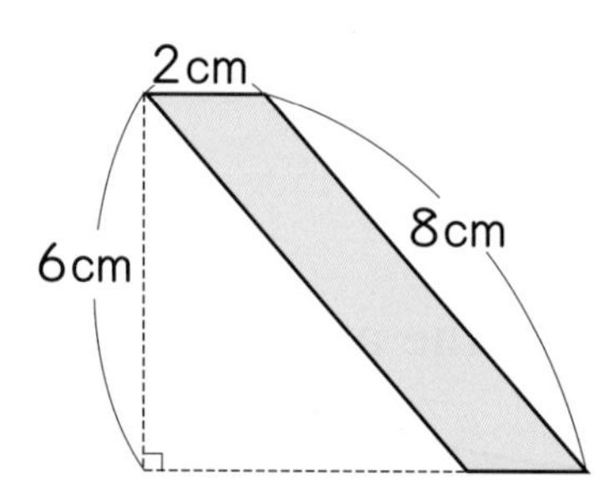


☐ **Can find the area of triangles by using the formula.**

**2** Let's find the area of the following triangles.

① 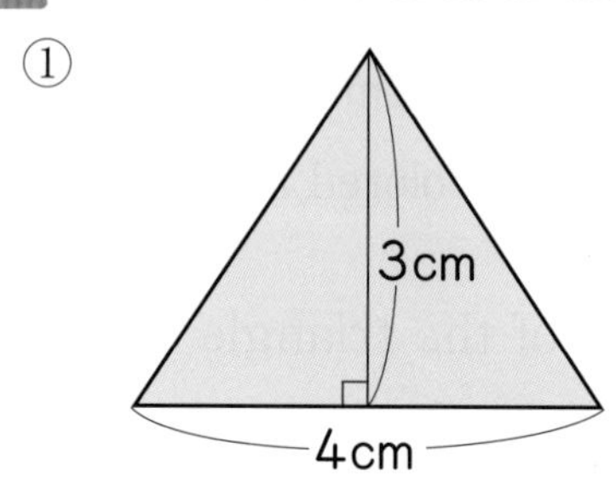


② 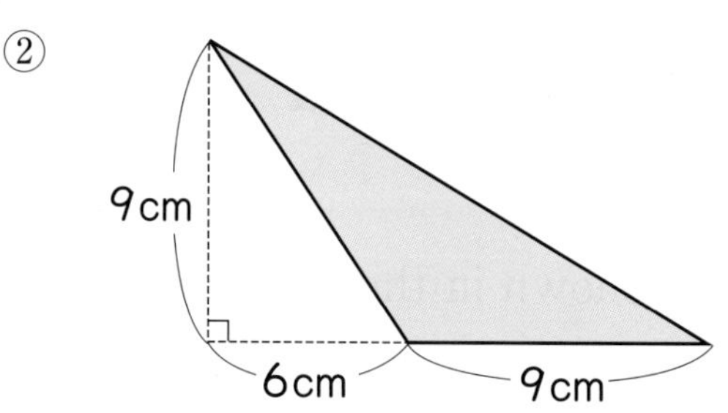


☐ **Can find the area of trapezoids by using the formula.**

**3** Let's find the area of the following trapezoids.

① 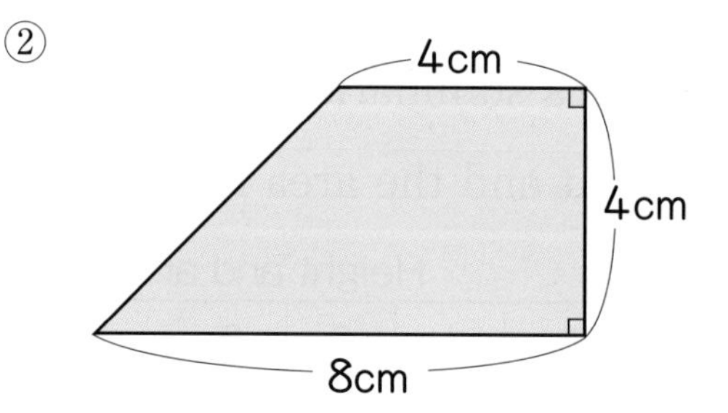


② 


☐ **Can find the area of rhombuses by using the formula.**

**4** Let's find the area of the following rhombuses.

① 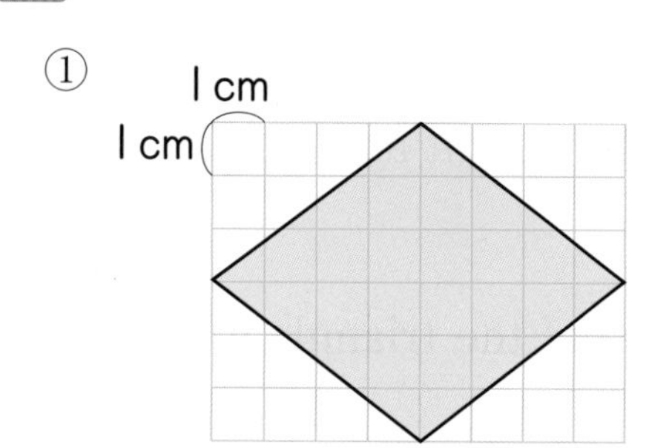


② 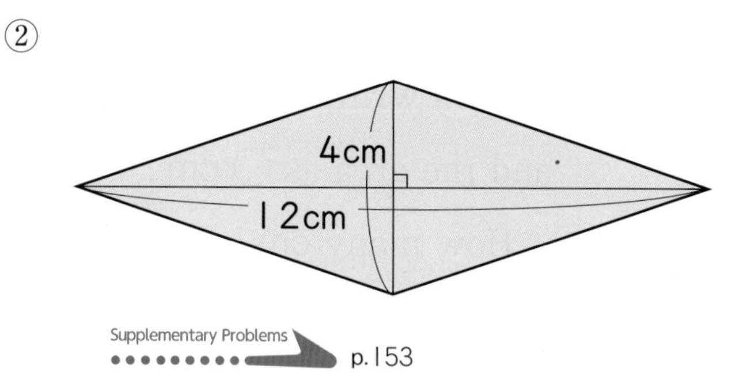


Supplementary Problems p.153

# Usefulness and efficiency of learning

**1** Let's find the area of the following figures.

Can find the area of figures by using the formulas.

① Parallelogram

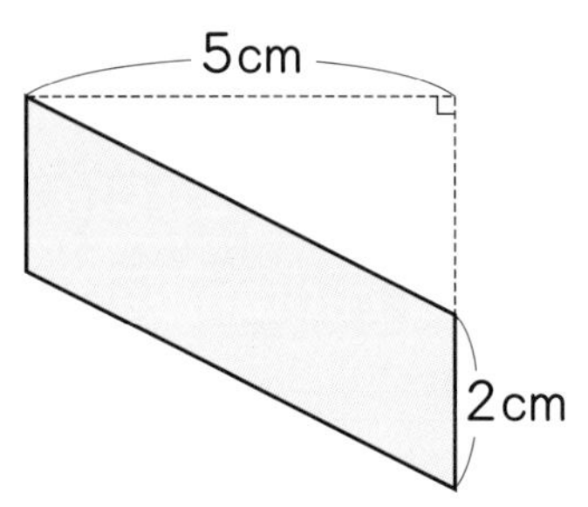


②

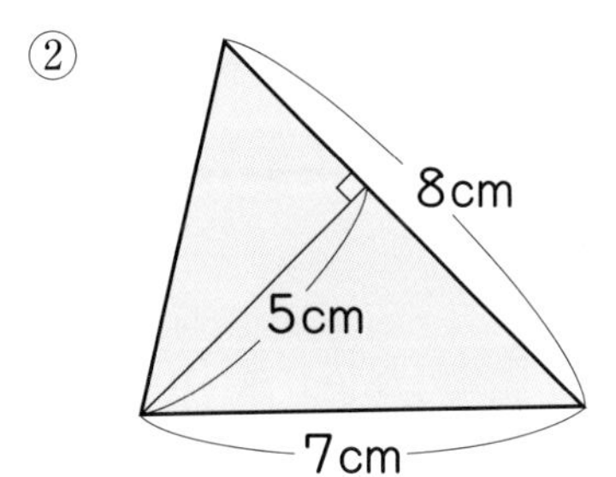


③ Trapezoid

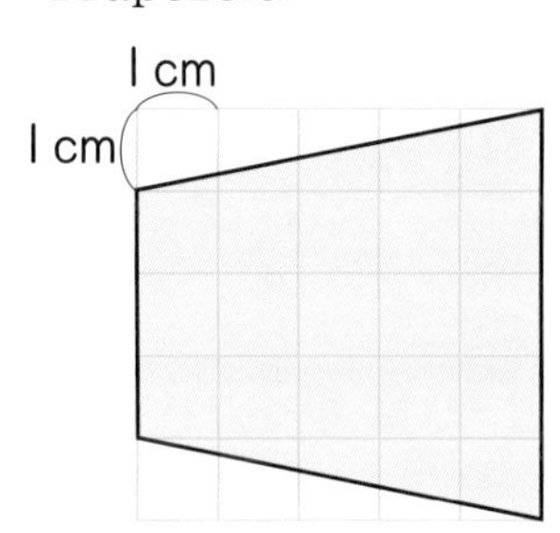


④

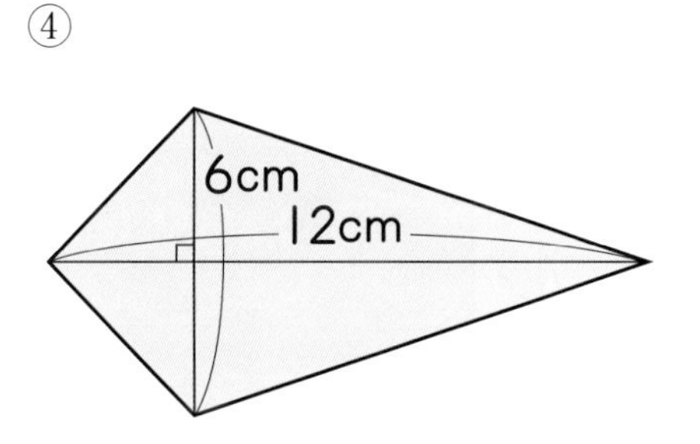


**2** Let's find the length of the base of the following figures.

Can find the length of the base by using the formulas.

① Parallelogram with an area of $216\ cm^2$.

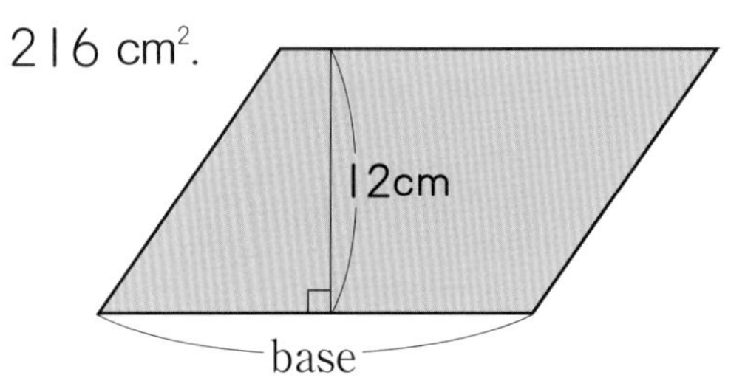


② Triangle with an area of $135cm^2$.

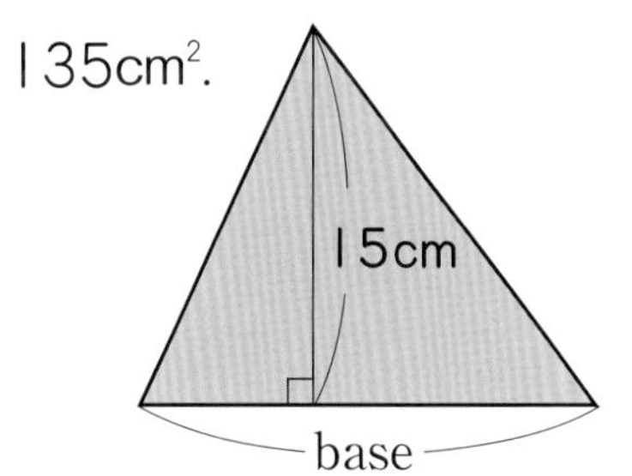


**3** Let's find the colored area of the following figures.

Can find the area of various figures.

①

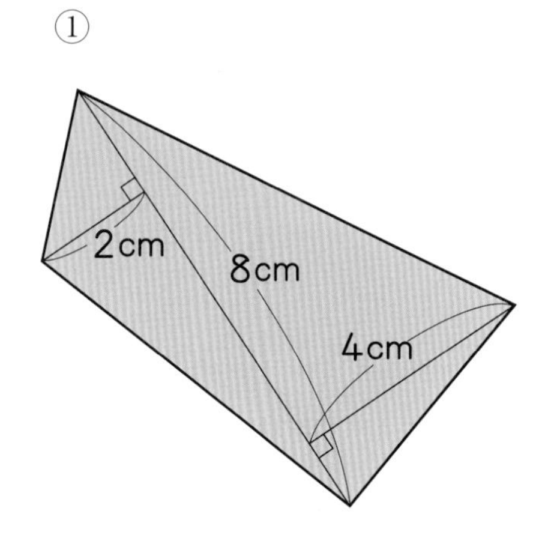


②

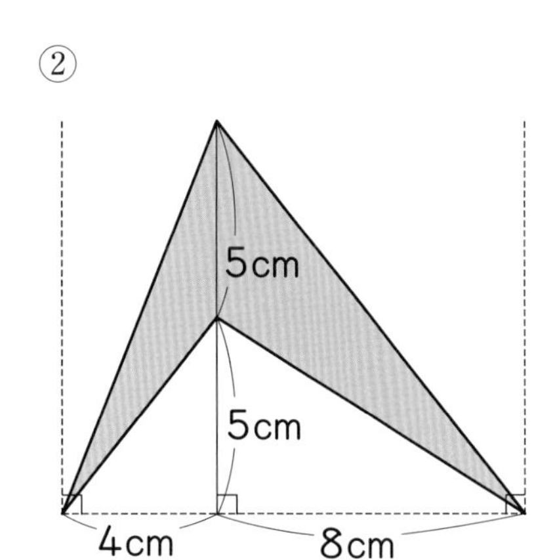


③

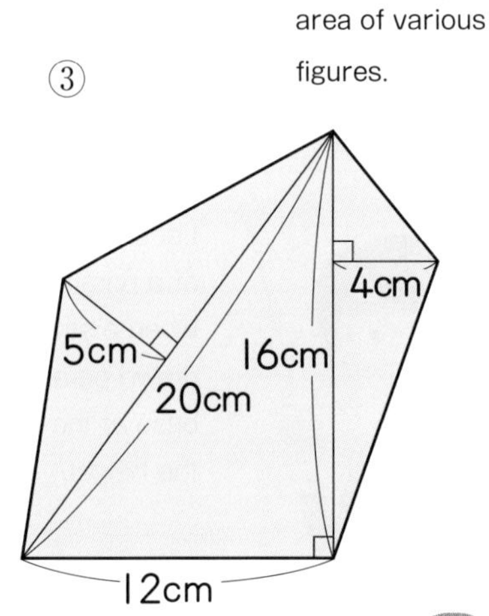

**Problem**

**Let's review the formulas for the area of figures.**

## Parallelogram?

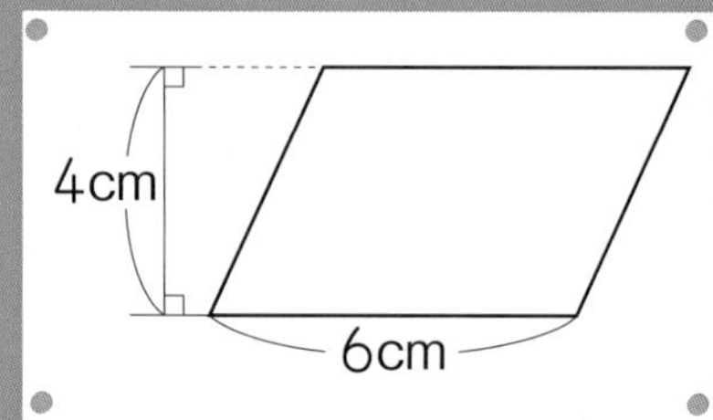


Area of a parallelogram = base × height

$6 \times 4 = 24$ 24cm²

If you think of changing it into a rectangle at the midpoint of the height,

$4 \times 6 = 24$ 24cm²

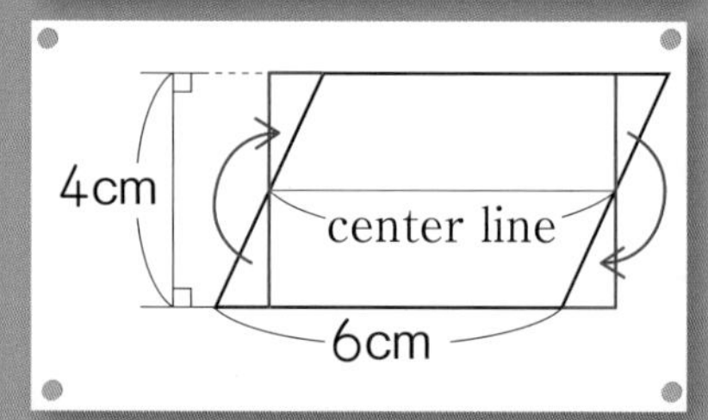


The straight line drawn parallel to the base at the midpoint of the height is called "center line."

Area of a parallelogram, **height × center line**

## Triangle?

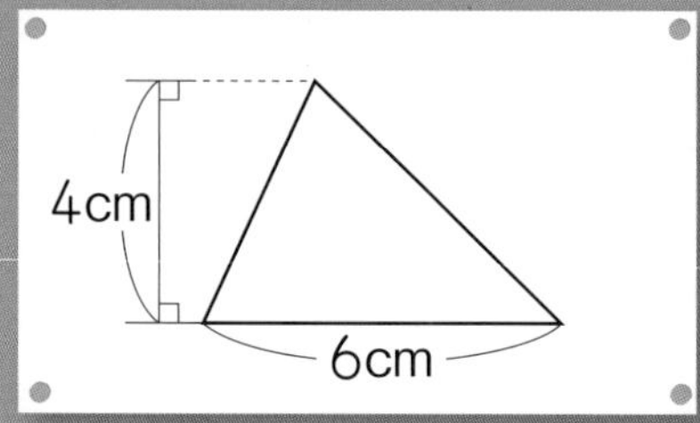


Area of a triangle = base × height ÷ 2

$6 \times 4 \div 2 = 12$ 12cm²

If you think of changing it into a rectangle at the midpoint of the height,

The length of the center line is base ÷ 2.

$4 \times 6 \div 2 = 12$ 12cm²

$4 \times 3 = 12$

Center line

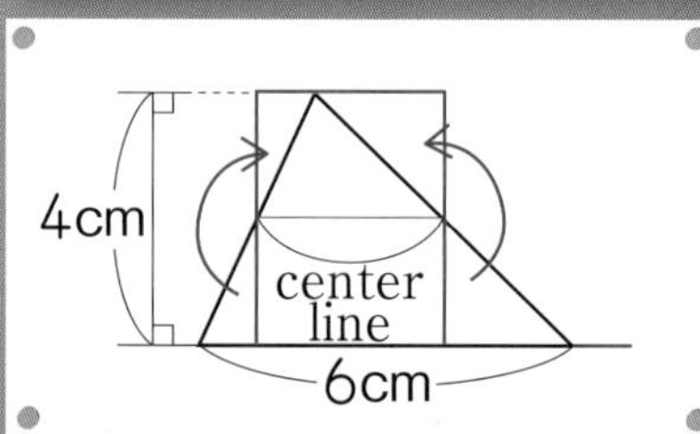


Area of a triangle, **height × center line**

I can think of it as a rectangle.
Hiroto

Does the area of a triangle also become height × center line?
Yui

## Trapezoid?

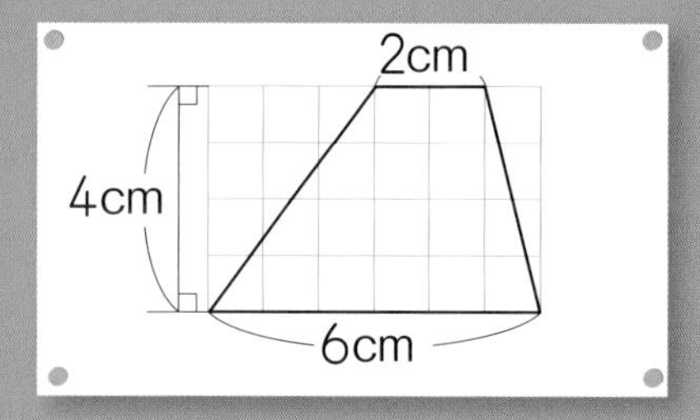


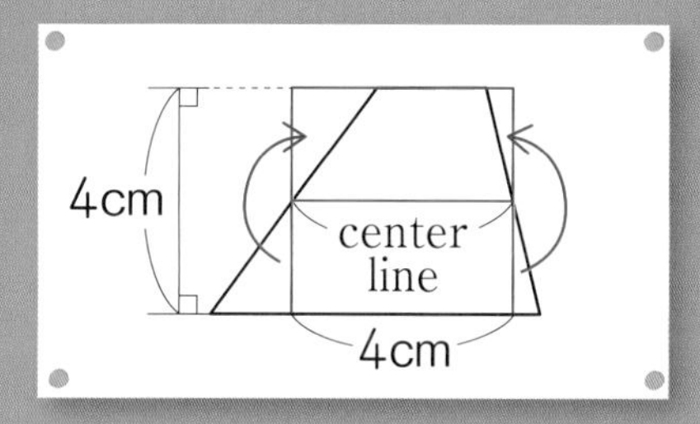


Area of a trapezoid $=$ (upper base $+$ lower base) $\times$ height $\div 2$

$(2+6)\times 4\div 2=16$　　<u>16cm²</u>

If you think of changing it into a rectangle at the midpoint of the height,

$4\times 4=16$　　**height × center line**

◎ You can also think from the formula of a trapezoid.

$$(2+6)\times 4\div 2=8\times 4\div 2$$
$$=4\times(8\div 2)$$
$$=4\times 4$$

The length of the center line is (upper base+ lower base) ÷ 2

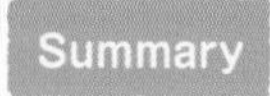


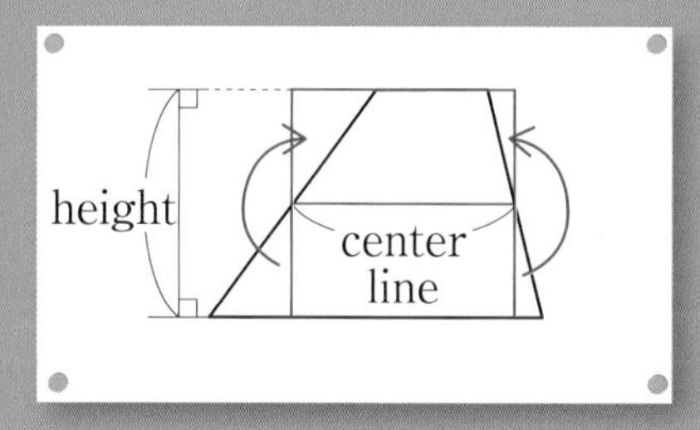


◎ Any figure will result in a rectangle in the same way.

◎ The area of a parallelogram, triangle, and trapezoid can be found by height × center line or center line × height.

In a trapezoid, when you look at the squares, you can understand the length of the center line.

Daiki

If I use the formula of a trapezoid...

Nanami

Can I think in the same way for rhombuses and other figures?

Yui

# What does the "÷ 2" do in a formula?

Want to explore Considering the area formula of a triangle

The formula to find the area of a triangle is "base × height ÷ 2." The method to find the area of a triangle was deduced from the following ways of thinking.

Nanami's idea

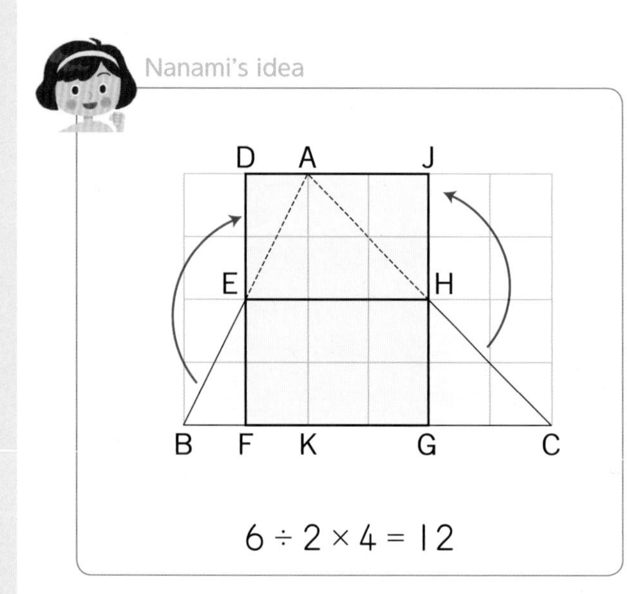


Hiroto's idea

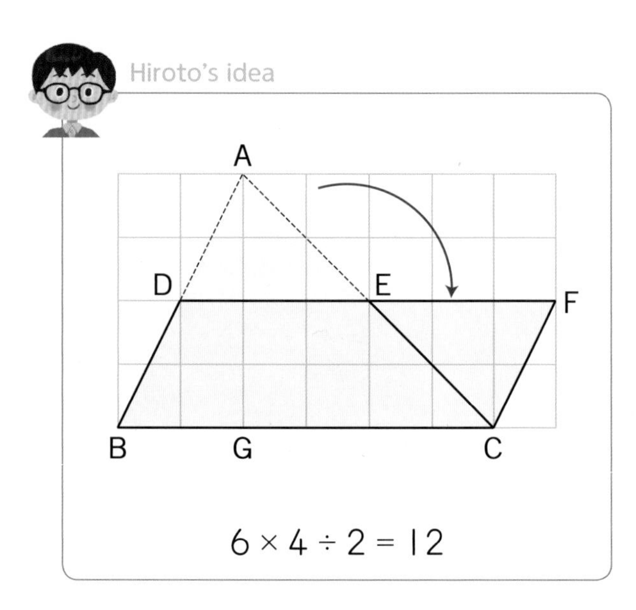


Yui's idea

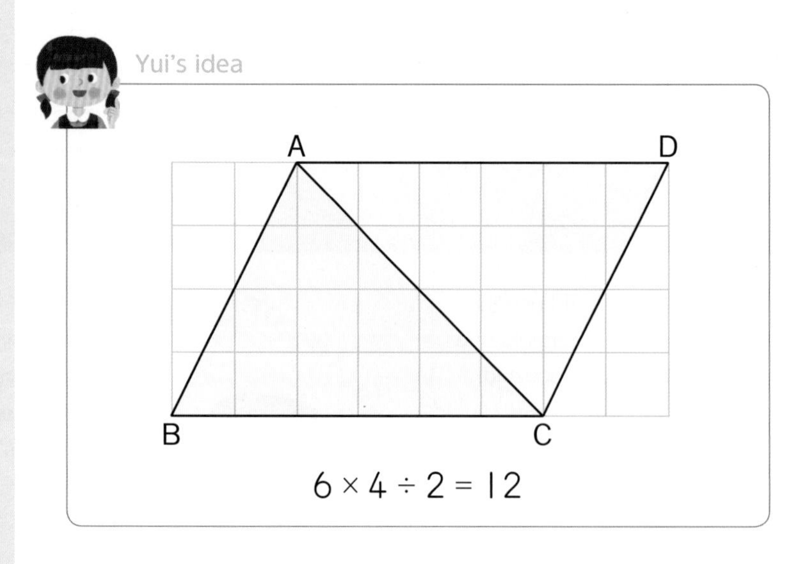

Think by myself

**1** What does the "÷ 2" do in the formula to find the area of a triangle?

Let's try to think by looking at the ways of thinking of the children.

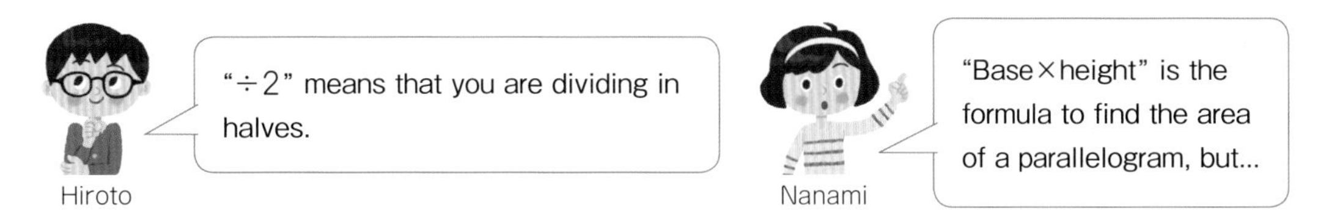


Think in groups

**2** Let's divide in groups. Explain to the classmates your own ideas about what the "÷ 2" does in the formula.

Also, let's hear the ideas of other classmates and discuss the similarities and differences with your own ideas. Make a summary.

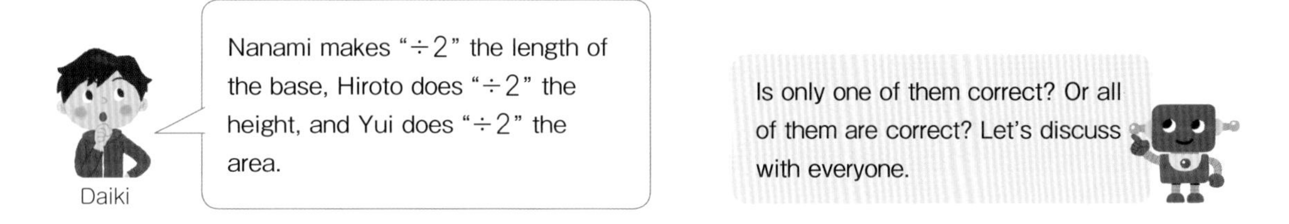

Nanami makes "÷2" the length of the base, Hiroto does "÷2" the height, and Yui does "÷2" the area.
Daiki
Is only one of them correct? Or all of them are correct? Let's discuss with everyone.

Think as a class

**3** Let's present what you summarized in your group.

Let's listen to other's opinions and try to summarize what you think.

Also, let's reflect on the ways of thinking of the three children on what the "÷ 2" does in the formula, and make a summary in your own words.

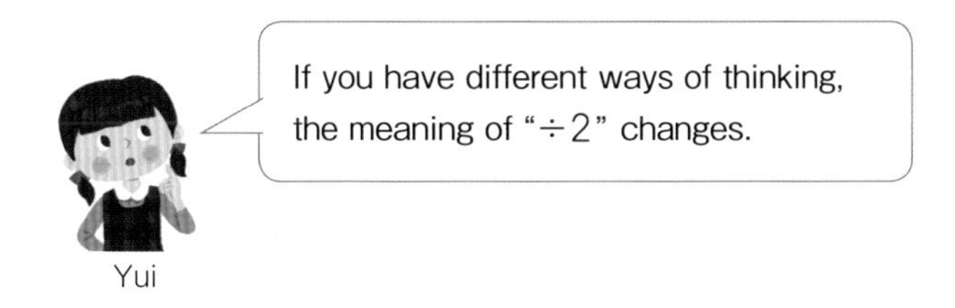

If you have different ways of thinking, the meaning of "÷2" changes.
Yui

01503

Find the ?

# How to make a beautiful polygon with origami paper?

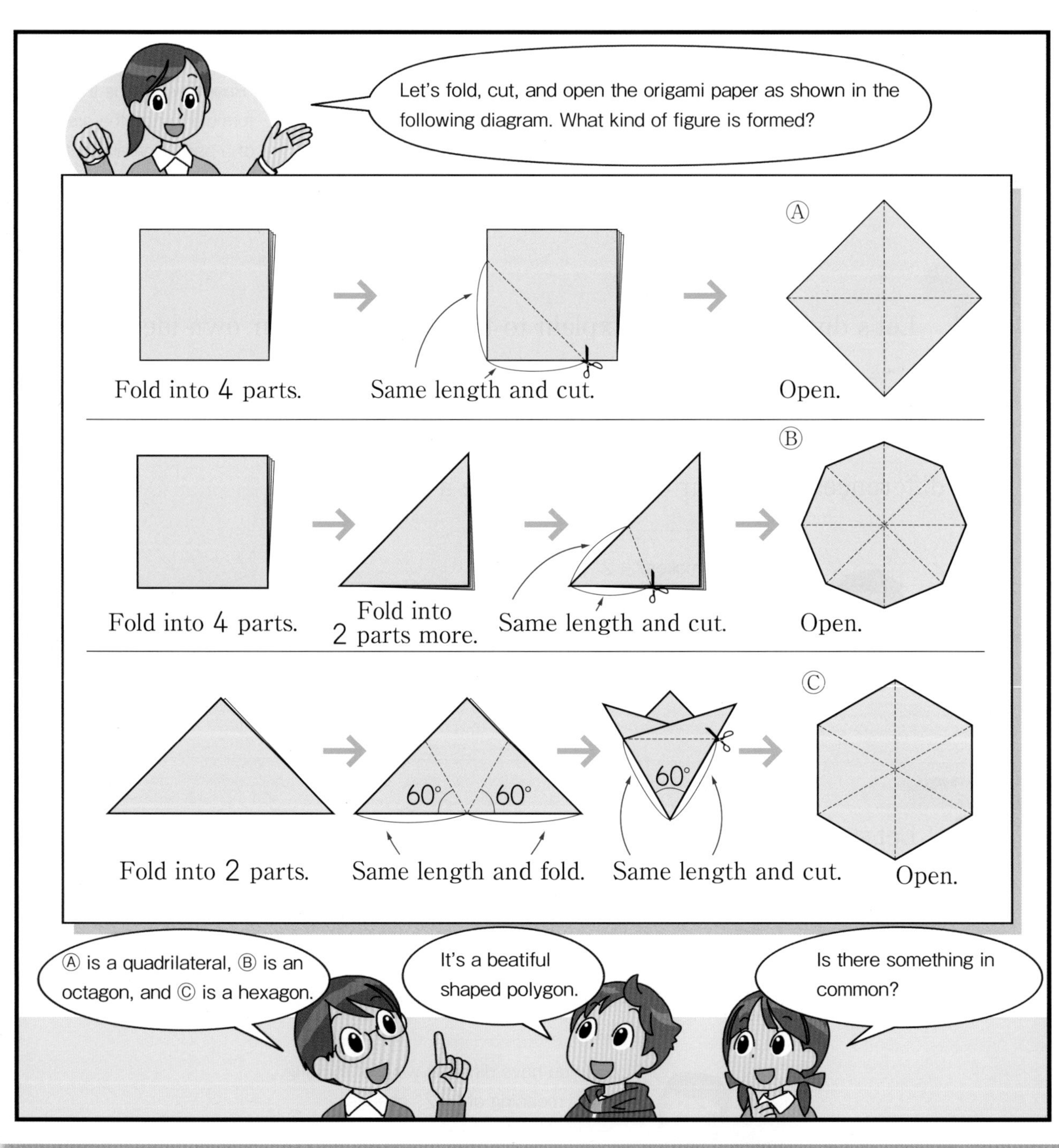
Let's fold, cut, and open the origami paper as shown in the following diagram. What kind of figure is formed?
Fold into 4 parts.
Same length and cut.
Ⓐ
Open.
Fold into 4 parts.
Fold into 2 parts more.
Same length and cut.
Ⓑ
Open.
Fold into 2 parts.
60°
60°
Same length and fold.
60°
Same length and cut.
Ⓒ
Open.
Ⓐ is a quadrilateral, Ⓑ is an octagon, and Ⓒ is a hexagon.
It's a beatiful shaped polygon.
Is there something in common?

**Problem** What kind of figure is a beautiful shaped polygon?

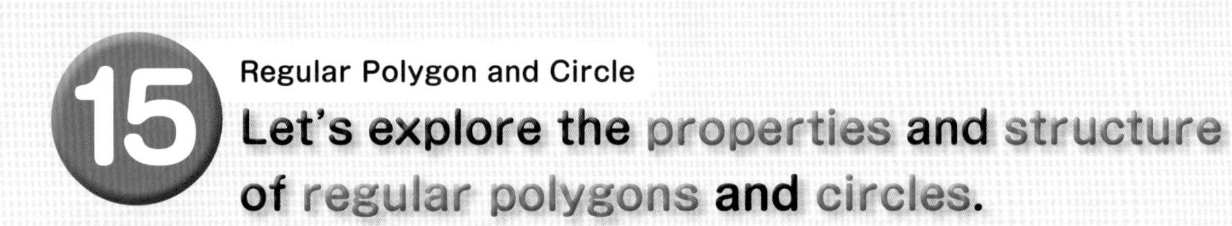

# 15 Regular Polygon and Circle

# Let's explore the properties and structure of regular polygons and circles.

## 1 Regular polygons

Want to explore

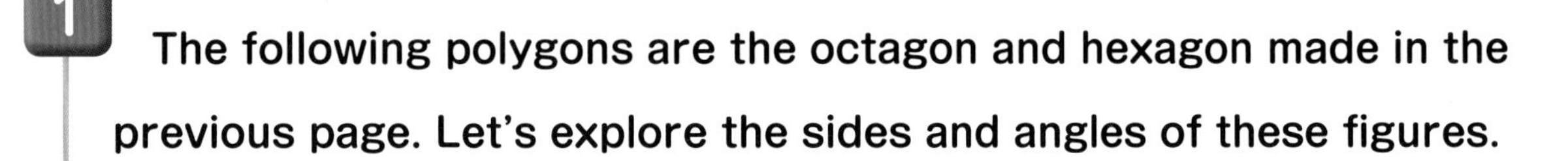

**1** The following polygons are the octagon and hexagon made in the previous page. Let's explore the sides and angles of these figures.

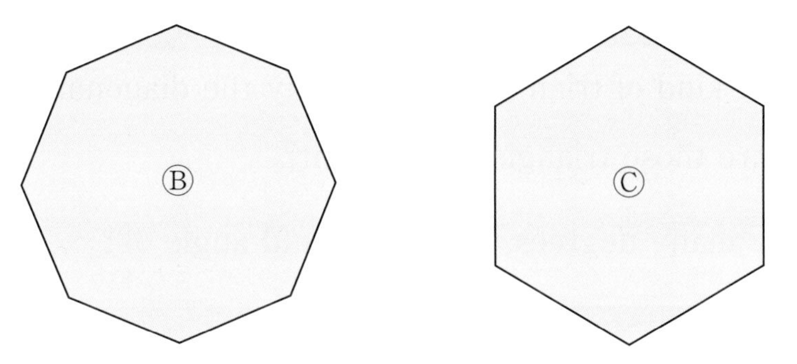


① How many sides and angles are there in each polygon?

② Let's try to measure the length of the sides and the size of the angles in each polygon.

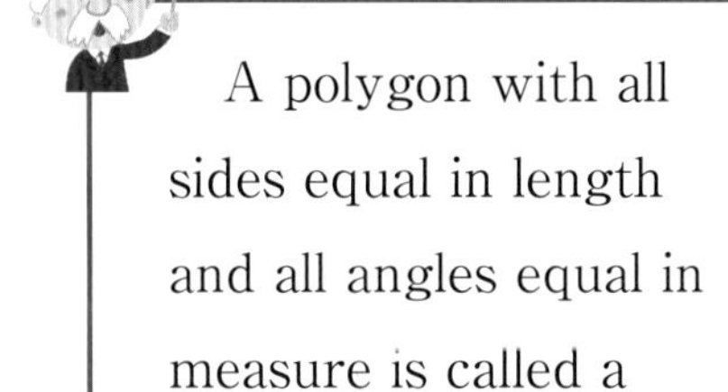

A polygon with all sides equal in length and all angles equal in measure is called a **regular polygon**.

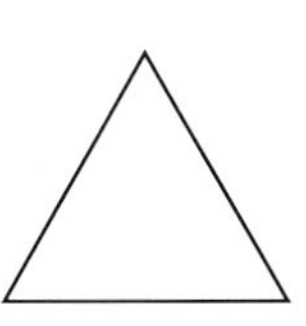
Regular triangle (Equilateral triangle)

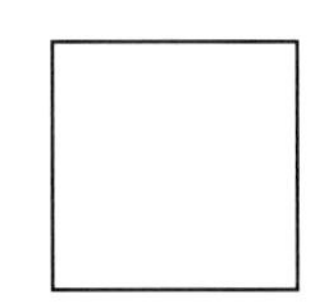
Regular quadrilateral (Square)

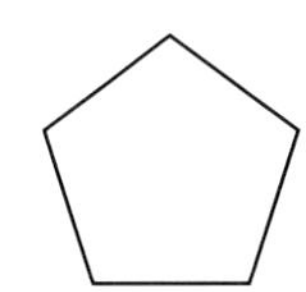
Regular pentagon

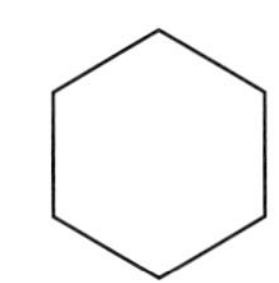
Regular hexagon

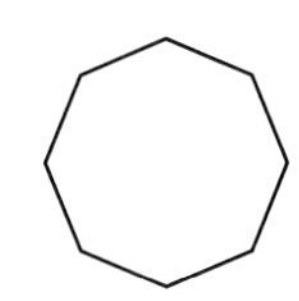
Regular octagon

Want to explore

Is the polygon shown on the right a regular polygon? Let's explain the reasons by using the words "length of the side" and "size of the angle."

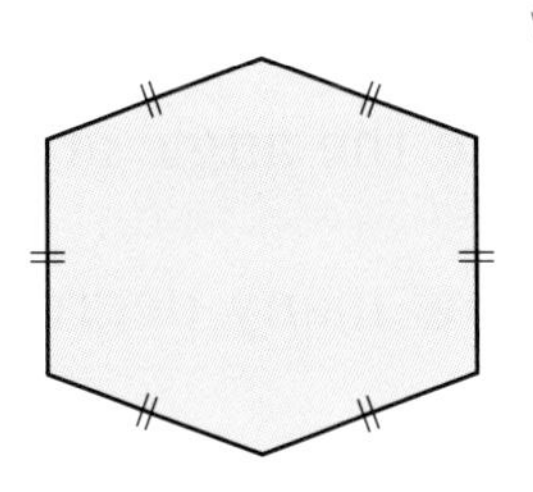

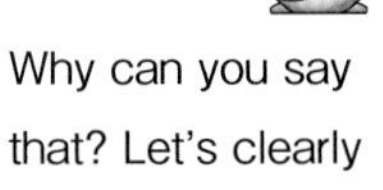

Way to see and think

Why can you say that? Let's clearly explain the reasons in order.

Want to know

## 2

**In the regular octagon shown on the right, if diagonals that connect opposite vertices are drawn, they will intersect at point O. By using this figure, let's explore the length of the sides and the size of the angles.**

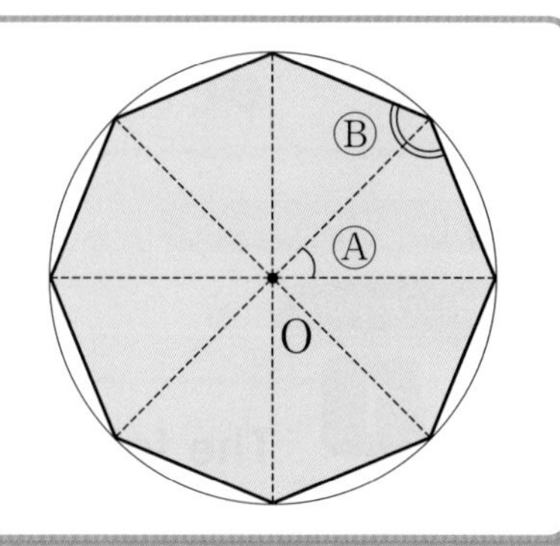


① Let's compare the length between point O and the vertices.

② What kind of triangle is formed by the diagonals?
Also, are those triangles congruent?

③ How many degrees is angle Ⓐ and angle Ⓑ?

In a regular polygon, the angles around the center of the circle are equally divided by the number of sides.

Want to confirm

**2** In the regular hexagon shown on the right, the diagonals connect opposite vertices and intersect at point O. Let's answer the following questions.

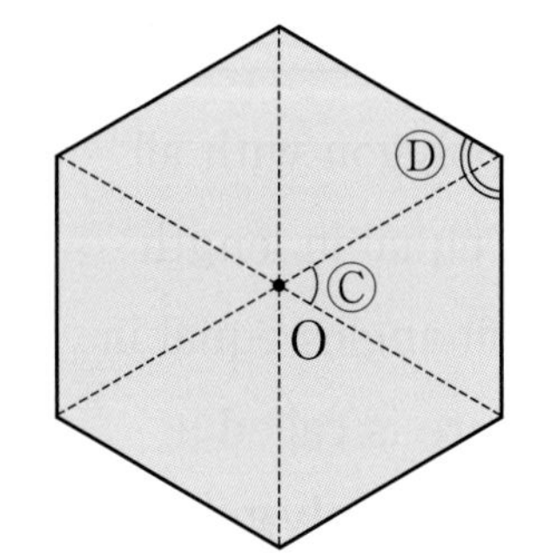


① What kind of triangle is formed by the diagonals?

② In how many equal parts do the diagonals divide the angle around point O?

③ How many degrees is angle Ⓒ and angle Ⓓ?

Activity

## 3 Nanami and Hiroto drew a regular octagon as shown below. Let's explain the ideas of them.

Nanami's idea

I decided the length of one side, and drew angles with a size of 135°.

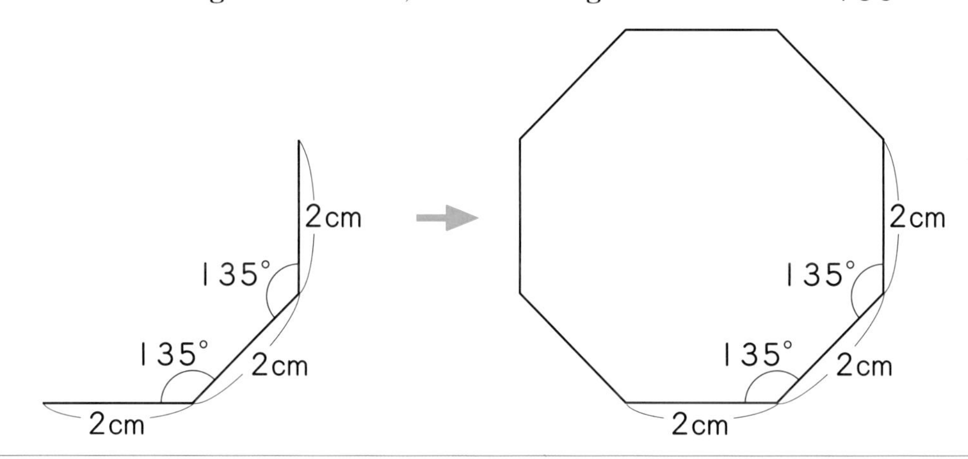


Hiroto's idea

Using a circle, I equally divided the angle around the center of the circle to draw.

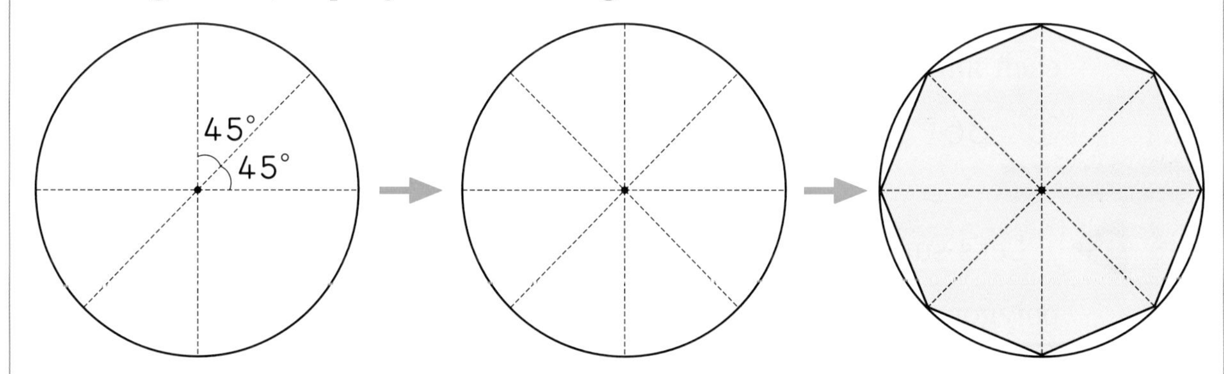


Want to try

Let's draw a regular pentagon by dividing the angle around the center of the circle into 5 equal parts.

① How many degrees should angle Ⓐ be?

② After drawing the regular pentagon, let's find the size of angles Ⓑ, Ⓒ, and Ⓓ.

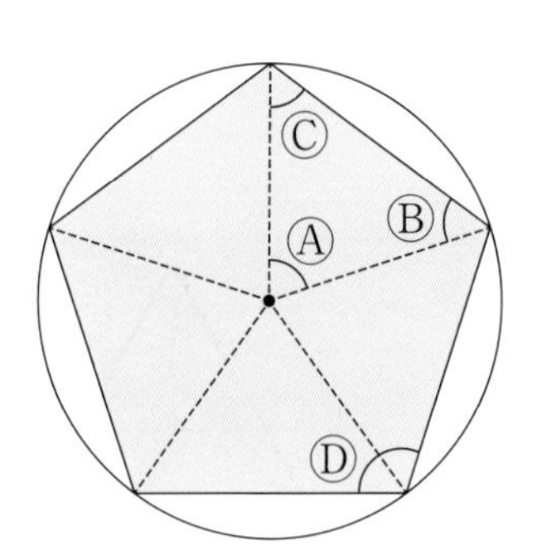

Want to think

## 4 Let's think about how to draw a regular hexagon by using a circle.

① Let's draw a regular hexagon by dividing the angle around the center of the circle into 6 equal parts.

② As shown in the following diagram, let's use a compass to separate the perimeter of the circle by the length of the radius.

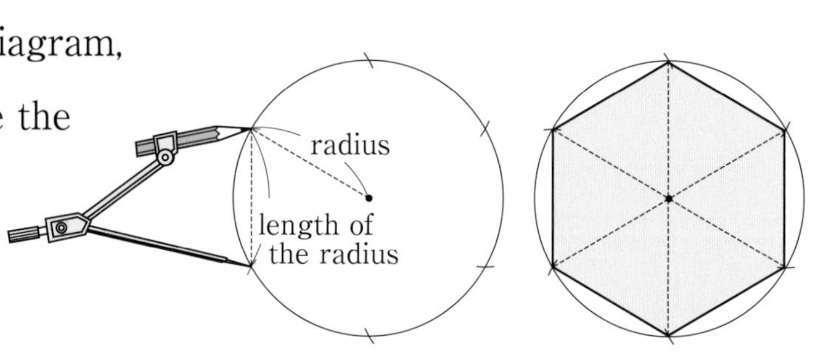


③ Let's explain the reasons why the regular hexagon can be drawn with the above method. Use the words "regular triangle (equilateral triangle)," "side," and "length of the radius."

Want to confirm

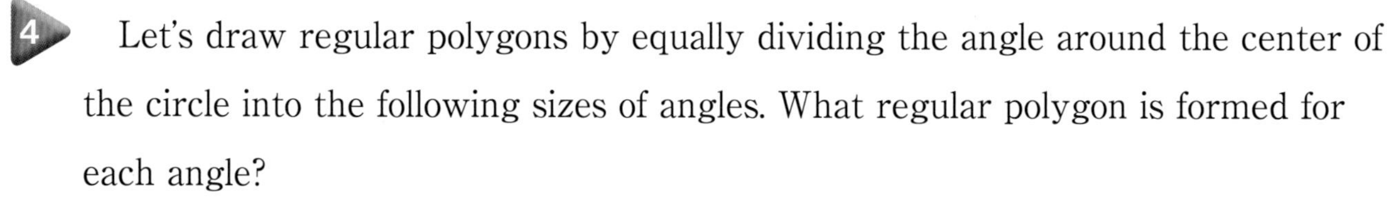

**4** Let's draw regular polygons by equally dividing the angle around the center of the circle into the following sizes of angles. What regular polygon is formed for each angle?

① 36° ② 90° ③ 40° ④ 60°

Want to summarize

**5** Let's summarize the number of sides and the size of angles of regular polygons.

| | Regular triangle | Regular quadrilateral (square) | Regular pentagon | Regular hexagon | Regular octagon |
|---|---|---|---|---|---|
| Number of sides | 3 | | | | |
| Size of angle Ⓐ | 120° | | | | |
| Size of angle Ⓑ | 60° | | | | |

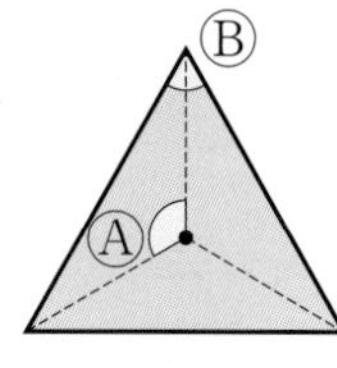

Regular triangle

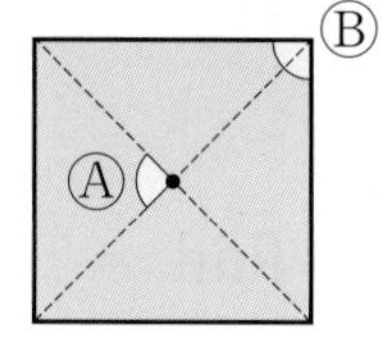

Regular quadrilateral (square)

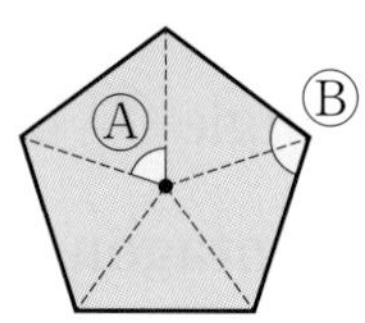

Regular pentagon

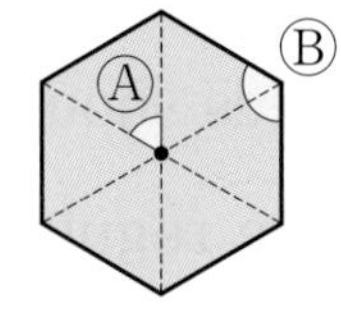

Regular hexagon

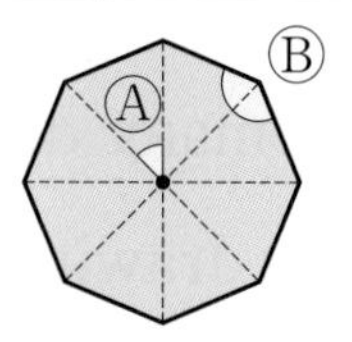

Regular octagon

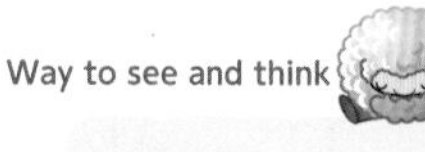


If you try to summarize, what kind of things do you understand?

## 2 Diameter and circumference

Want to know Relationship between circumference and diameter

**1** **A regular hexagon which fits perfectly into a circle with a 4 cm diameter and a square into which a circle with a 4 cm diameter fits perfectly were drawn. Let's explore the relationship between the diameter and the perimeter of the circle.**

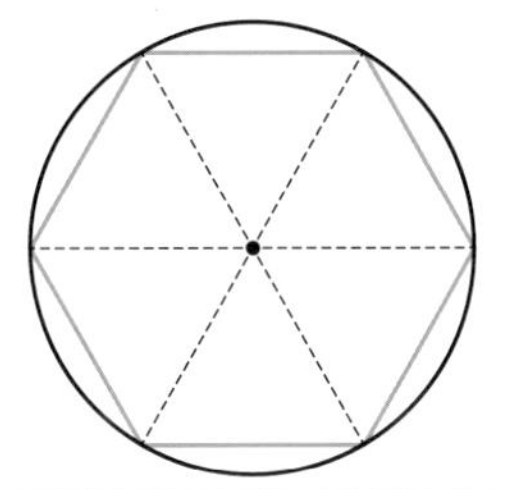

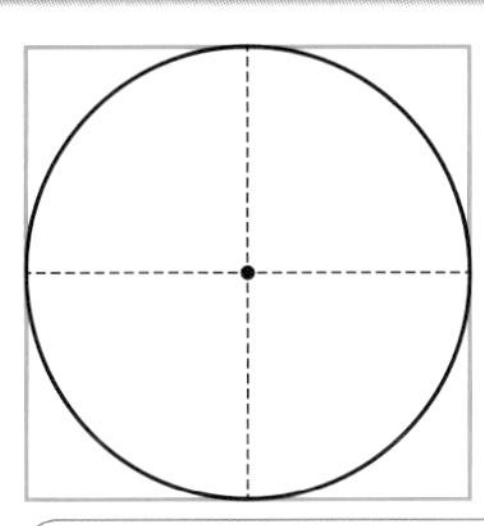

The perimeter of the circle is likely to increase as the diameter increases.

What kind of relationship is there?

Purpose Is there any relationship between the perimeter of the circle and the diameter?

The perimeter of a circle is called **circumference**. A line that bends like a circumference is called a **curve**.

Want to find

① How many times the diameter of the circle is the perimeter of the regular hexagon?

② How many times the diameter of the circle is the perimeter of the square?

③ Let's compare the perimeter of the regular hexagon and the square with the circumference. What kind of things do you understand? Let's write inequality signs in the □.

diameter × 3 □ circumference □ diameter × 4

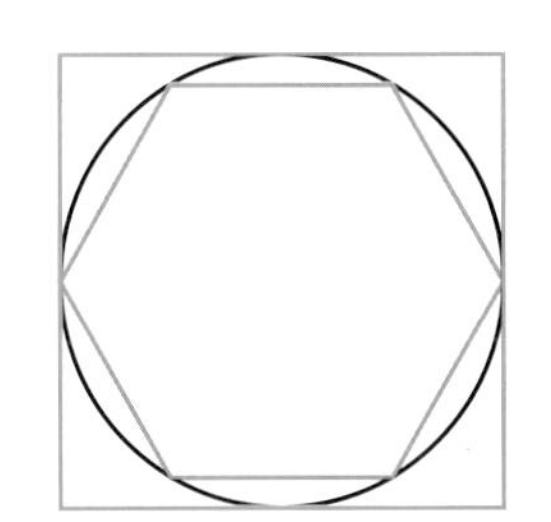

Summary

The circumference is longer than 3 times the diameter and shorter than 4 times the diameter.

## 2

**Let's cut the circles Ⓐ, Ⓑ, and Ⓒ from page 163 which have diameters of 4 cm, 8 cm, and 12 cm respectively. Also, cut the circumference rulers from page 166 and 167. Let's try to explore the circumferences by how many cm each circle advances in one rotation.**

Nanami

If the diameter becomes larger, then the circumference is likely to be longer.

When the diameter is 4cm, the circumference becomes between 12 cm and 16 cm.

Hiroto

**Purpose** What kind of relationship is there between the circumference and diameter?

① Let's write the results in the following table.

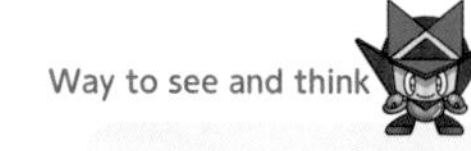


Can I use a method to find rules between two quantities changing together?

| | Circle Ⓐ | Circle Ⓑ | Circle Ⓒ | |
|---|---|---|---|---|
| Circumference (cm) | | | | |
| Diameter (cm) | 4 | 8 | 12 | |

② Let's estimate how many cm a circle with a 16 cm diameter will advance in one rotation.

③ Let's cut circle Ⓓ with a 16 cm diameter from page 161, and confirm how many cm the circle will advance in one rotation.

④ When the diameter increases 2 times, 3 times, or 4 times, how does the circumference change?

⑤ Are the circumference and the diameter proportional?

⑥ What do we need to know in order to find the circumference of various diameters?

⑦ Approximately, how many times of the diameter is the circumference?

Let's find the nearest hundredths place round number by rounding off the thousandths place.

| | Circle Ⓐ | Circle Ⓑ | Circle Ⓒ | Circle Ⓓ |
|---|---|---|---|---|
| Circumference (cm) | | | | |
| Diameter (cm) | 4 | | 12 | |
| Circumference ÷ diameter | | | | |

**Summary**

Regardless of the circle's size, circumference ÷ diameter is always the same number.

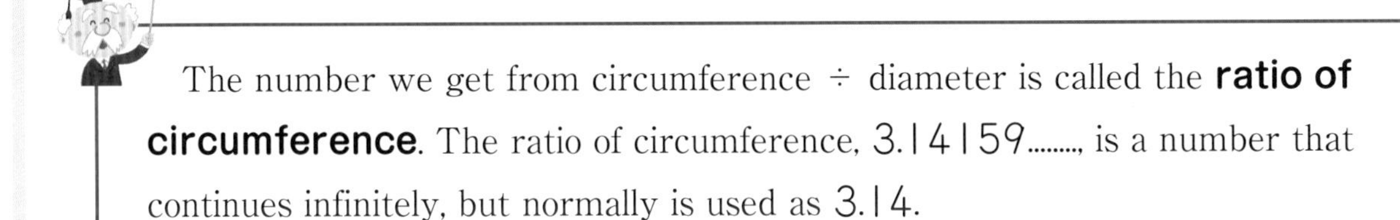

The number we get from circumference ÷ diameter is called the **ratio of circumference**. The ratio of circumference, 3.14159........, is a number that continues infinitely, but normally is used as 3.14.

**Ratio of circumference = circumference ÷ diameter**

Want to confirm

**1** Let's explore the relationship between the circumference and the diameter with circles of various sizes from our surroundings.

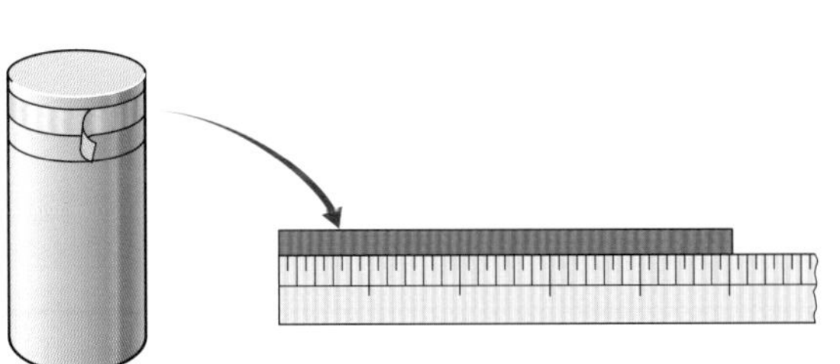

Measure the length of the circumference.

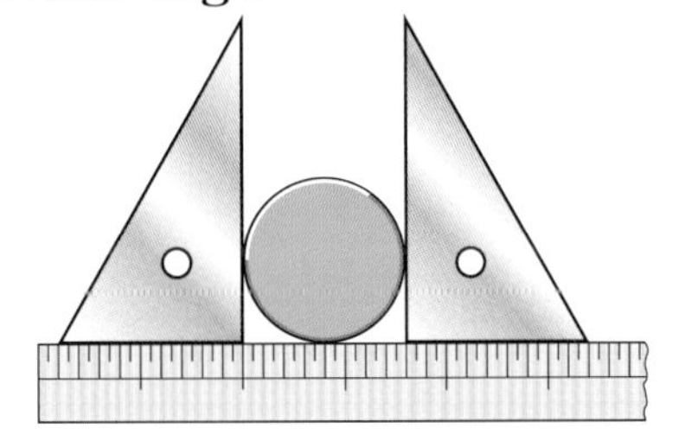

Measure the length of the diameter.

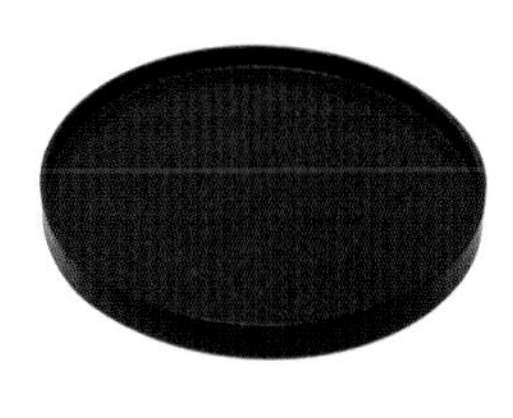

| | Can | | | | | | |
|---|---|---|---|---|---|---|---|
| Circumference (cm) | | | | | | | |
| Diameter (cm) | | | | | | | |

Want to think How to find the circumference

**How many m is the circumference of a circle with a diameter of 12 m?**

Monument (Shizuoka City, Shizuoka Prefecture)

You can find the circumference with the following formula.

**Circumference = diameter × 3.14**

Want to confirm

Let's find the circumference.

① Circle with a diameter of 15 cm.
② Circle with a radius of 25 m.
③ Circle with a radius of 32.5 cm.

Want to try

In the playground, a circle with a radius of 10 m was drawn. How many meters is the circumference?

Yui

If the radius is 10 m, the diameter is...

Want to think How to find the diameter

The length around the can shown on the picture is 62.8 cm. How many cm is the diameter of this can?

① Let's consider the diameter as □ cm, and write the number that applies in the math sentence "circumference = diameter × 3.14."

② How many cm is the diameter of the can?

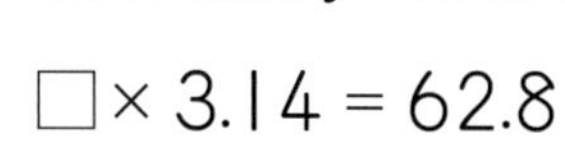

□× 3.14 = 62.8

Want to confirm

**5** Let's find the diameter of the circles with the following lengths as circumference.

① 28.26 m ② 31.4 cm ③ 37.68 cm

Want to try

**6** The photo on the right shows a tire used in a large truck in Kurume City, Fukuoka Prefecture. The circumference of the tire is 12.62 m. Let's find the diameter of this tire. Answer with the nearest hundredths place round number by rounding off the thousandths place.

Tire (Kurume City, Fukuoka Prefecture)

## That's it

### How many meters is the diameter of the cherry tree?

6 children can outstretch their hands to surround the cherry tree shown on the right. Approximately, how many meters is the diameter of this tree?

Let's find the answer by assuming that the length of each child's outstretched arms is 1.5 m and the ratio of circumference is 3 instead of 3.14.

Daigo-Sakura (Maniwa City, Okayama Prefecture)

Let's estimate the length.

## History of the ratio of circumference

The ratio of circumference is represented as a decimal number, 3.14159265358979......, which continues infinitely. The number has been computed now to 10 trillion decimal places by a supercomputer, but it was very difficult to calculate in ancient times.

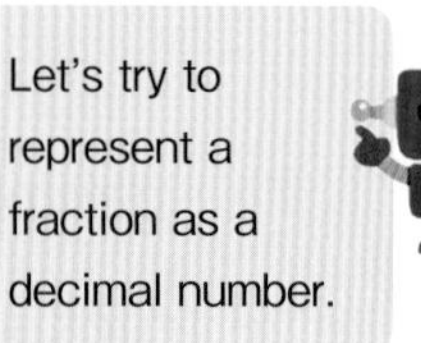

Let's try to represent a fraction as a decimal number.


(1) Many years ago, 3 was used as the ratio of circumference.

(2) About 4000 years ago, $3\frac{1}{8}$ and $3\frac{13}{81}$ were used as the ratio of circumference in Egypt and other countries.

(3) About 2000 years ago, Archimedes in Greece found that the ratio of circumference is between $3\frac{10}{71}$ and $3\frac{1}{7}$.

Archimedes

(4) In China about 1500 years ago, Zu Chongzhi used the fractions $\frac{22}{7}$ and $\frac{355}{113}$.

Zu Chongzhi

$\frac{22}{7} = \square$ $\frac{355}{113} = \square$

(5) In Japan about 300 years ago, Seki Takakazu found that the ratio of circumference is slightly smaller than 3.14159265359 by calculating.

Seki Takakazu

# What you can do now

**Can draw regular polygons.**

**1** Let's draw regular polygons by using circles.

① Regular hexagon

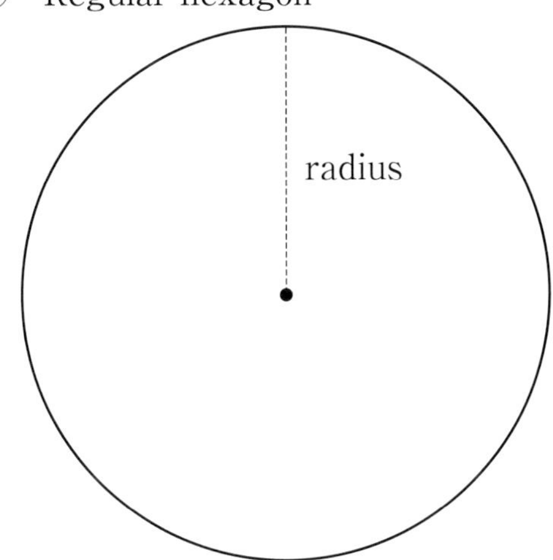


② Regular pentagon

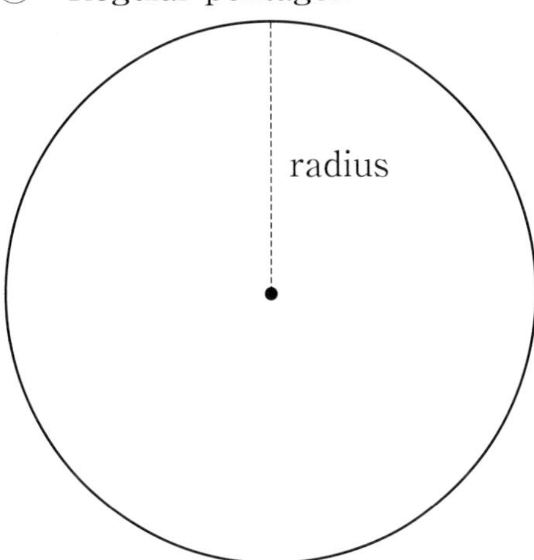


**Understanding the properties of regular polygons.**

**2** Let's draw regular polygons by equally dividing the angle around the center of the circle into the following sizes of angles. What regular polygon is formed for each angle?

① 30° ② 40° ③ 120°

**Understanding the relationship between the circumference and diameter of a circle.**

**3** Let's find the circumference of the following circles.

① Circle with a diameter of 6 cm. ② Circle with a radius of 5 cm.

**Understanding the relationship between the circumference and diameter of a circle.**

**4** Let's find the length of the diameters of the following circles.

① Circle with a circumference of 6.28 cm. ② Circle with a circumference of 12.56 cm.

Supplementary Problems p.154

# Usefulness and efficiency of learning

**1** Let's write the numbers that apply in each ☐ about the regular nonagon shown on the right.

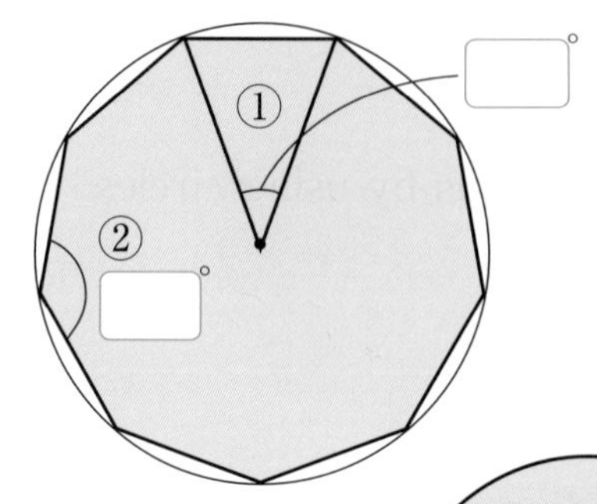


☐ Understanding the properties of regular polygons.

**2** Is the circumference of the circle proportional to its radius? Let's explore by writing the numbers that apply in the following table.

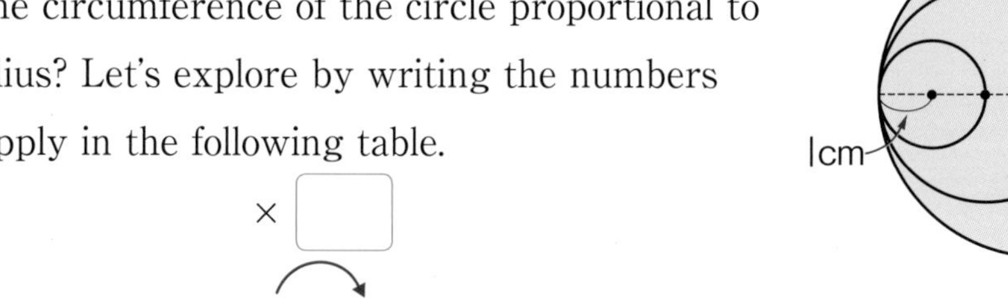


☐ Understanding the relationship between the circumference and radius of a circle.

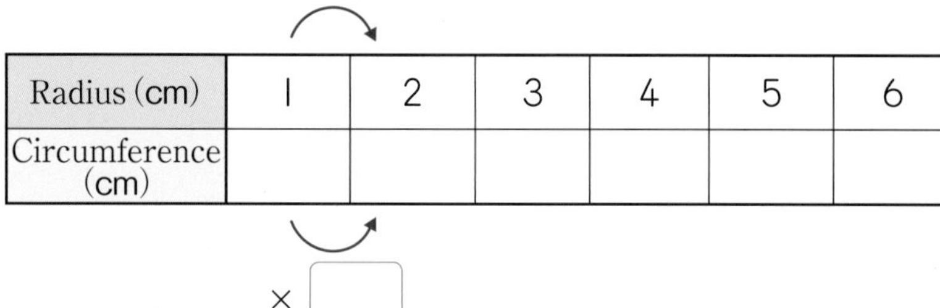

× ☐

| Radius (cm) | 1 | 2 | 3 | 4 | 5 | 6 |
|---|---|---|---|---|---|---|
| Circumference (cm) | | | | | | |

× ☐

**3** Let's find the perimeter of Ⓐ and Ⓑ, and compare them.

☐ Can find the circumference.

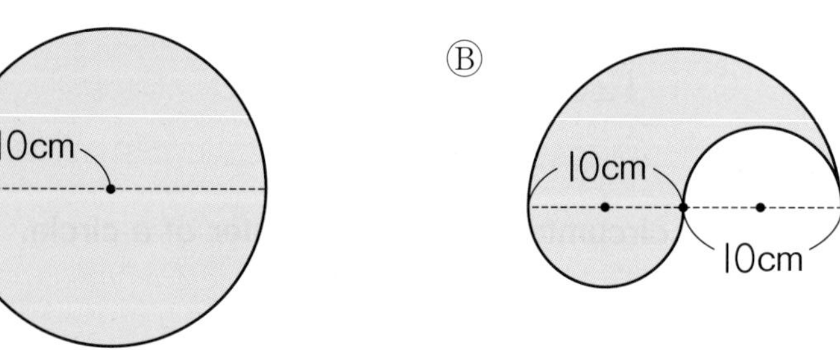


☐ Understanding the relationship between the circumference and diameter of a circle.

**4** As shown below, circle Ⓑ was drawn around circle Ⓐ. The radius of circle Ⓑ is 1 cm longer than the radius of circle Ⓐ, which is 2 cm. How many cm is the circumference of circle Ⓑ longer than the circumference of circle Ⓐ?

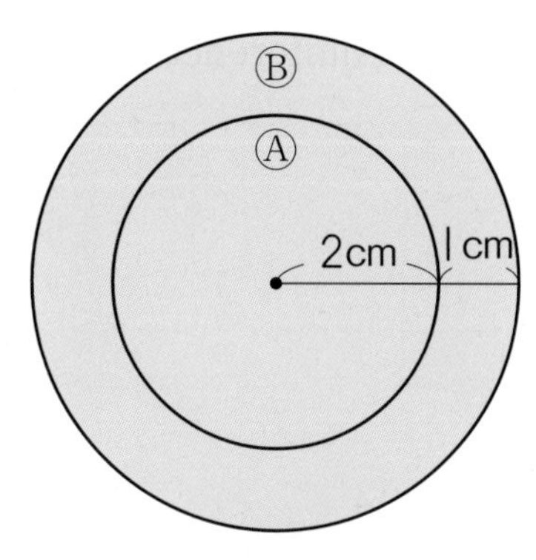


**Let's deepen.**

Is there something in our surroundings that uses the properties of circles?

Utilize in life.
Deepen.

# What is the relationship between the back wheel and training wheels of a bicycle?

**Want to find**

**A bicycle has training wheels. The diameter of the back wheel is 40 cm, and the diameter of the training wheels is 10 cm.**

**What kind of movement do the back wheel and training wheels have?**

Back wheel
40 cm
diameter
Training wheel
10 cm diameter
The wheels move together even though the sizes are different.
Nanami

① The back wheel turned once and moved from A to B as shown on the diagram below. Which of the following shows the movement of the center O of the back wheel? Choose one from Ⓐ to Ⓒ.

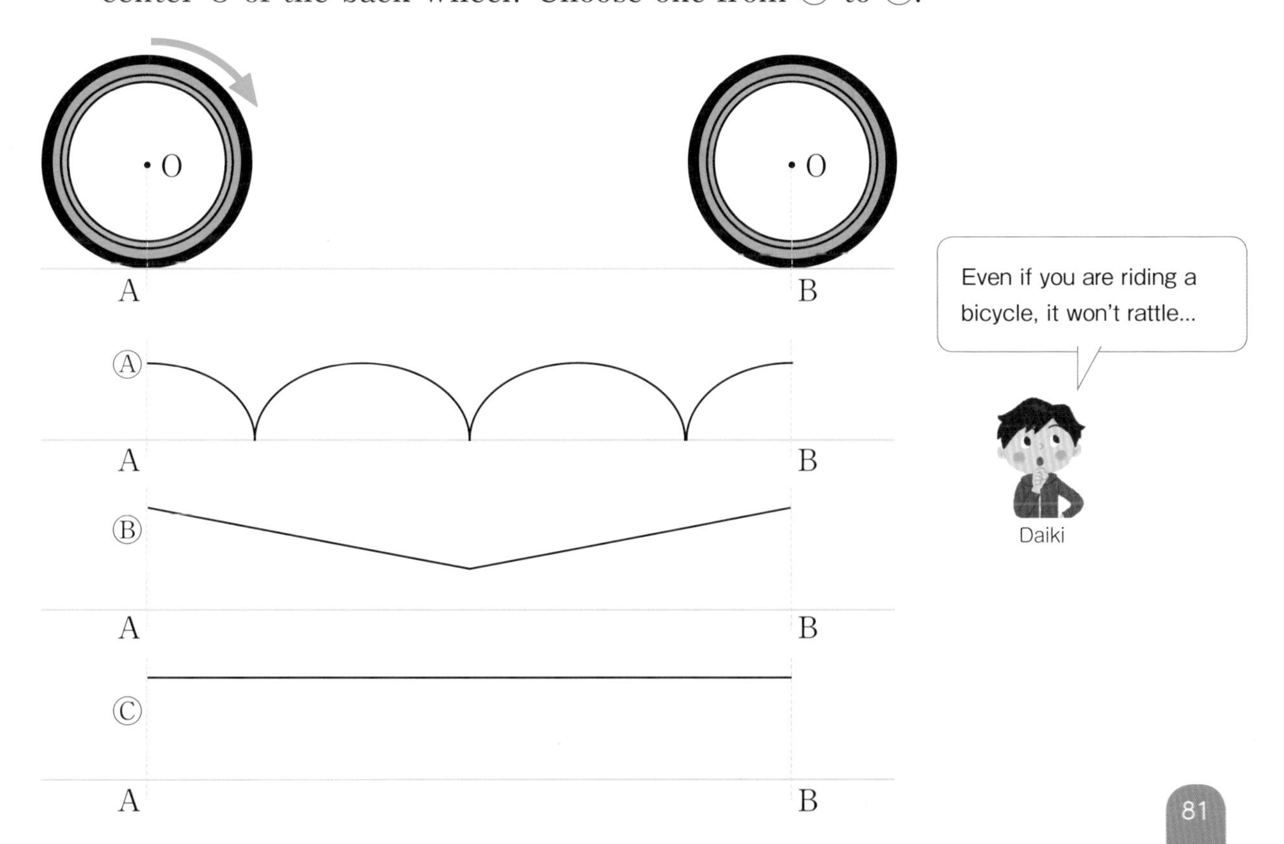
O
O
A
B
Ⓐ
A
B
Ⓑ
A
B
Ⓒ
A
B
Even if you are riding a bicycle, it won't rattle...
Daiki

② How many cm does the bicycle move when the back wheel turns once?

③ When the back wheel turns 12 times, how many times do the training wheels turn? Yui and Hiroto have the following ideas.

Consider that when the training wheels turn, they don't rise from the ground.

Yui's idea

First, find the length that the back wheel covered by turning 12 times. Then, divide the length by the circumference of the training wheels.

Hiroto's idea

When you think about the relationship between the diameters of the back wheel and training wheels, you can find the answer without the length of the circumference.

Which math expression did Yui and Hiroto use to find the answer? Let's choose one for each from Ⓐ to Ⓓ.

Ⓐ $12 \times 3$

Ⓑ $(40 \times 3.14 \times 12) \div (10 \times 3.14)$

Ⓒ $12 \times (40 \div 10)$

Ⓓ $40 \times 3.14 \times 3$

④ Let's write the reason why Hiroto can find the answer with his idea.

Want to deepen | Develop in High School

**Let's explore how point C moves when point C is considered as the point of the back wheel touching the ground at first.**

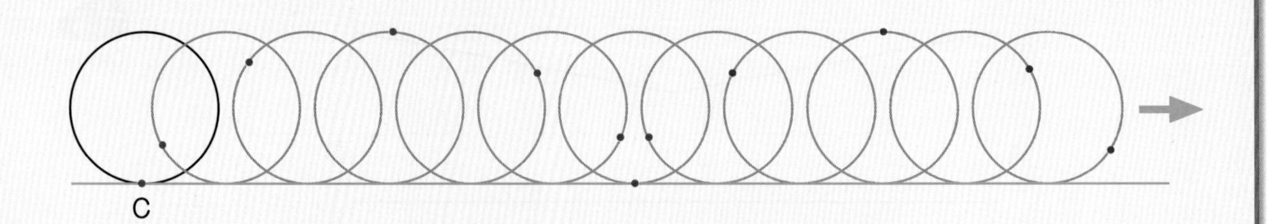

Find
the

# Who has the largest box?

Everyone made a box by drawing the net of a cuboid or cube.

1

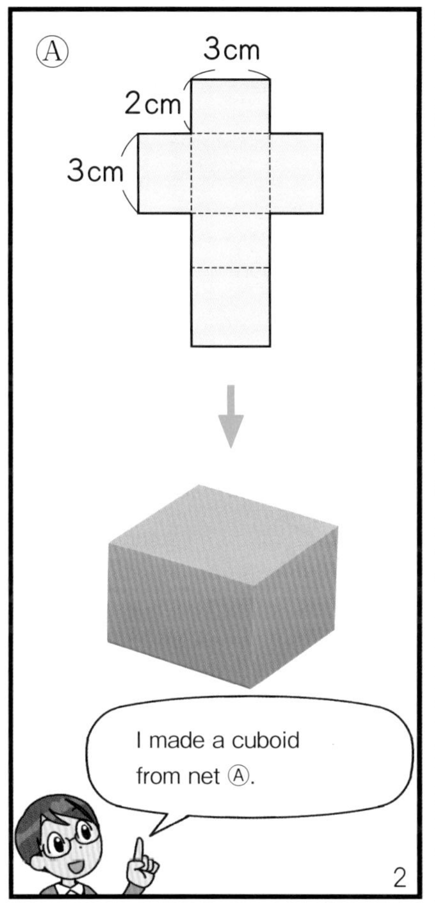
Ⓐ
3cm
2cm
3cm
I made a cuboid from net Ⓐ.
2

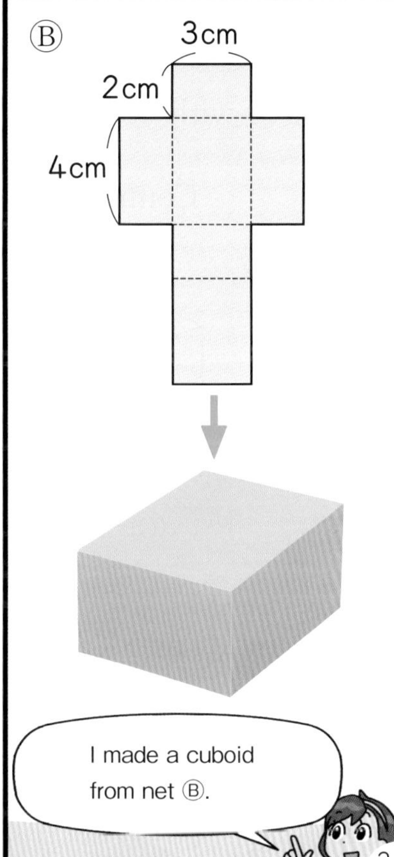
Ⓑ
3cm
2cm
4cm
I made a cuboid from net Ⓑ.
3

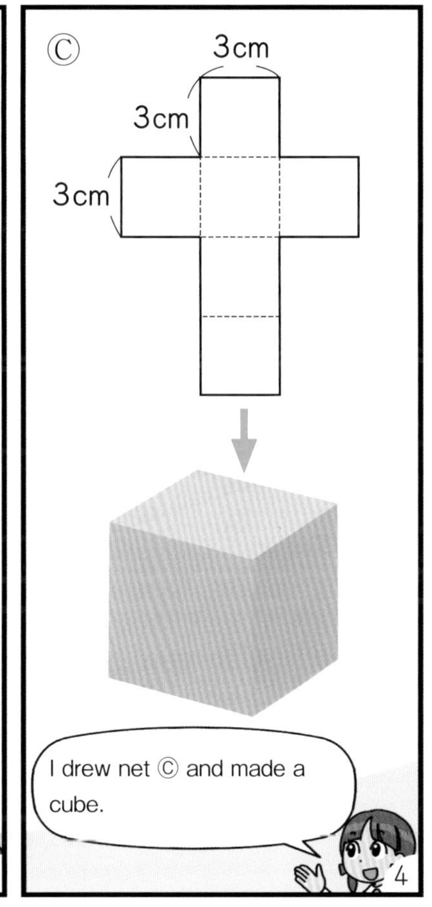
Ⓒ
3cm
3cm
3cm
I drew net Ⓒ and made a cube.
4

Who has the largest box?
Should we understand by arranging the boxes?

Let's use the nets in page 169, and actually make and compare the boxes.

5

Problem

How can we compare the sizes of the boxes?

# Volume
# Let's explore the size of cuboids and cubes, and how to find it.

## 1 Volume

Want to compare The size of cuboids and cubes

**1** **Let's compare the size of the cuboids and cube created in the previous page.**

① Let's compare the size of Ⓐ with Ⓑ and Ⓐ with Ⓒ.

Comparing Ⓐ and Ⓑ.

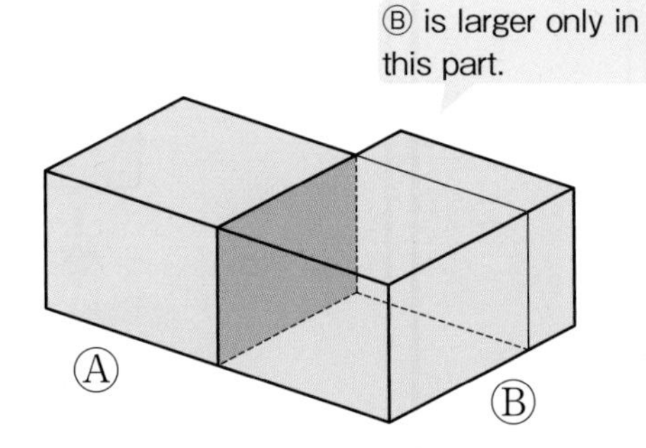

Ⓑ is larger only in this part.
Ⓐ
Ⓑ

Comparing Ⓐ and Ⓒ.

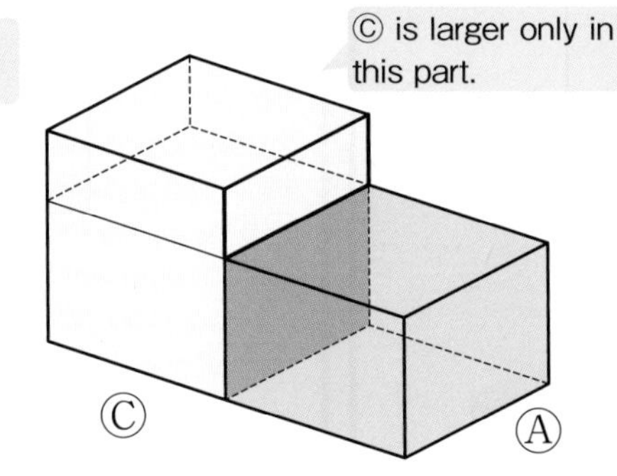

Ⓒ is larger only in this part.
Ⓒ
Ⓐ

Way to see and think

Thinking about aligning the height and width of the cuboids and cube.

② Can you compare the size of Ⓑ and Ⓒ?

Comparing Ⓑ and Ⓒ.

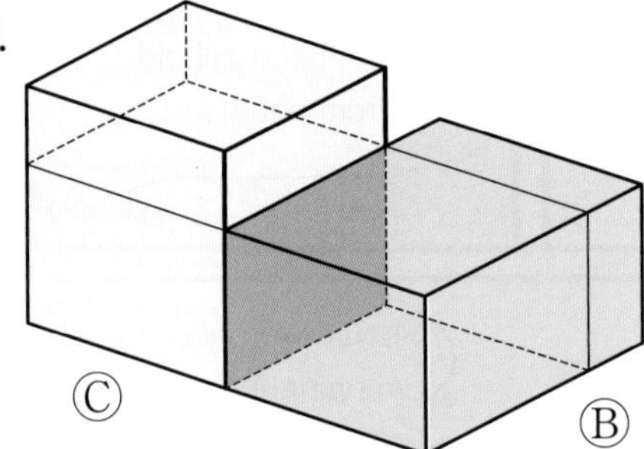

Ⓒ
Ⓑ

Daiki

We used squares with a side of 1 cm to compare the area.

To compare the size of a cuboid or cube...

Yui

Purpose

How should we compare the size of a cuboid or cube?

Want to improve

③ We made the same solids by using 1 cm cubic blocks. Let's compare the number of cubic blocks needed to make boxes Ⓑ and Ⓒ.

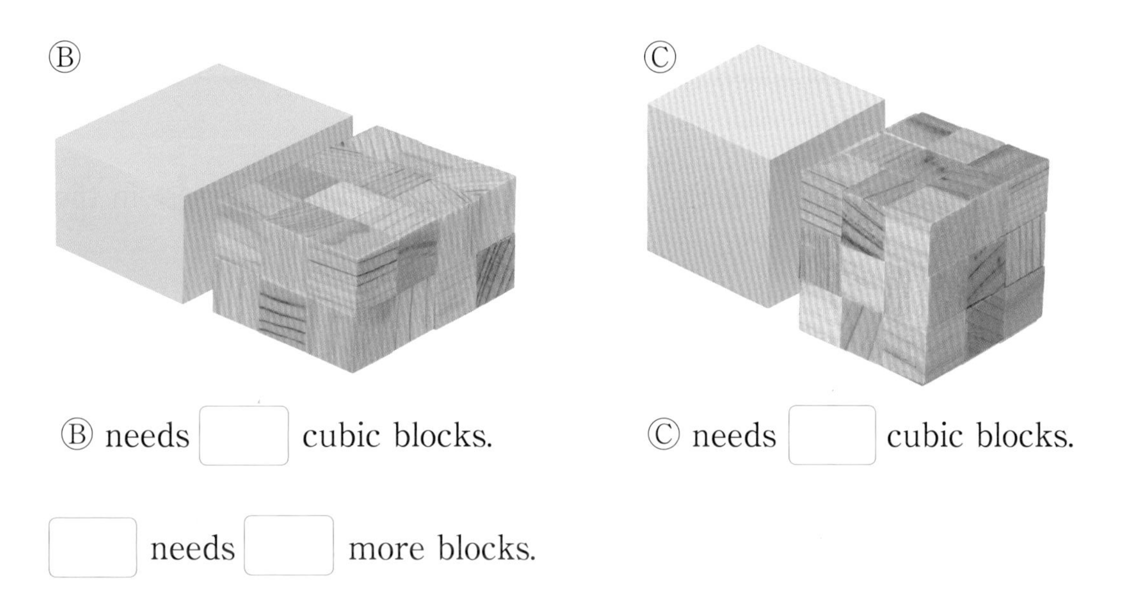


Ⓑ needs ☐ cubic blocks. Ⓒ needs ☐ cubic blocks.

☐ needs ☐ more blocks.

Summary

The size of a cuboid or cube can be represented by the number of cubic units. The size of one unit is a cube with a side of 1 cm.

Want to confirm

1 The following cuboids and cube were made by using 1 cm cubic units. How many 1 cm cubic units is the size of each solid?

①
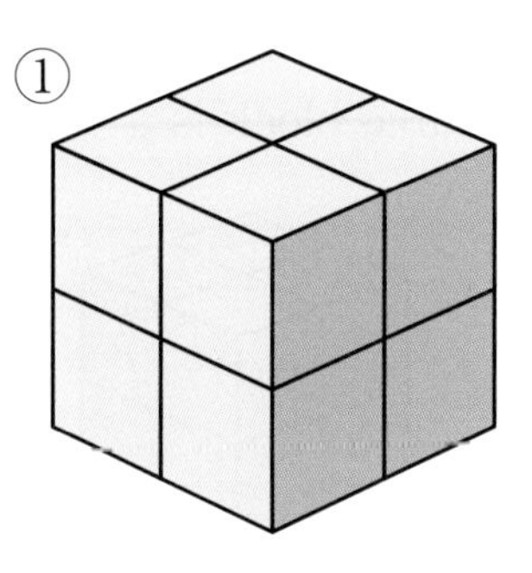

②
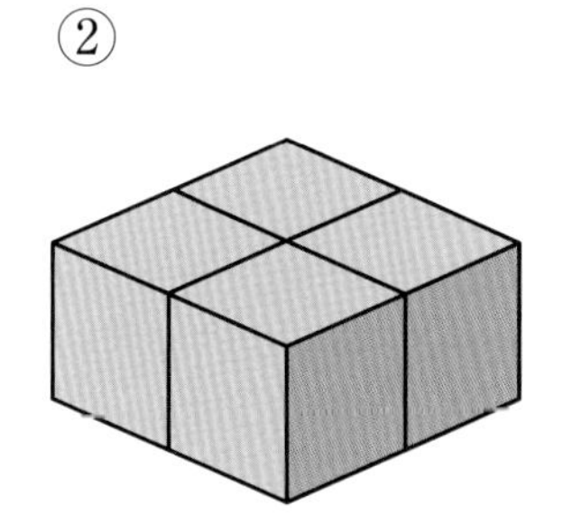

③
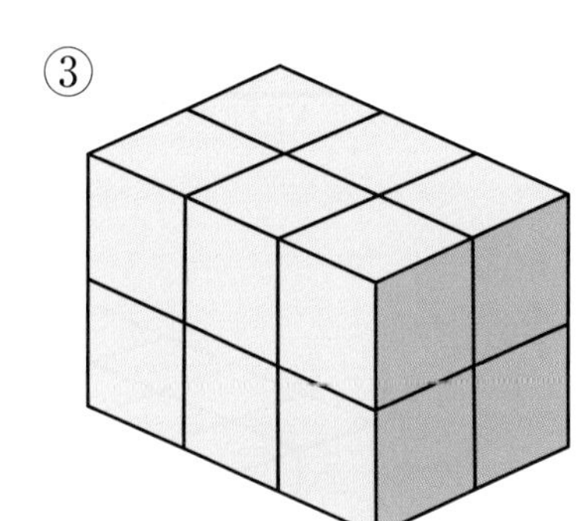

The size of a solid expressed by the number of units is called **volume**. The volume of a cube with sides of 1 cm is called 1 **cubic centimeter** (cc) and is written as 1 $cm^3$.

The unit $cm^3$ is a unit of volume.

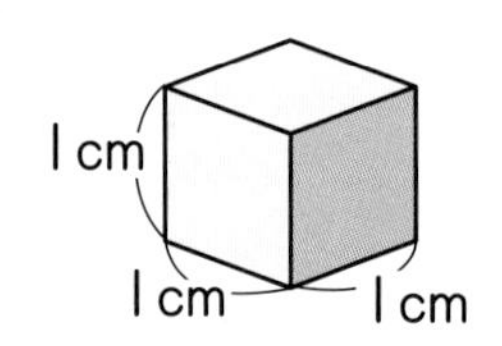


Want to represent

How many $cm^3$ is the volume of ①, ②, and ③ from ▶1 in the previous page?

Want to try

Let's find the volume of the following cuboid and cube.

①

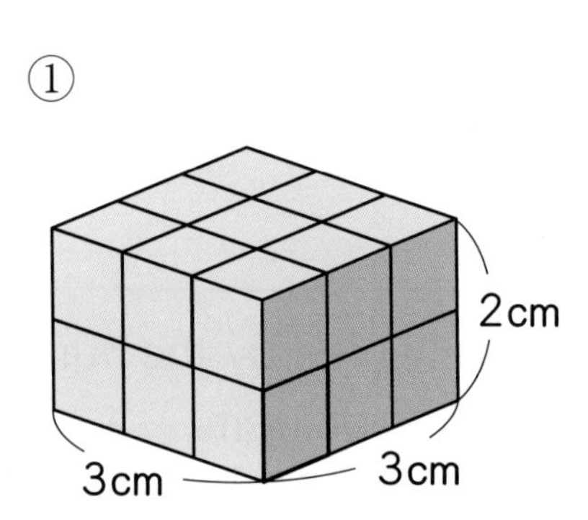


②

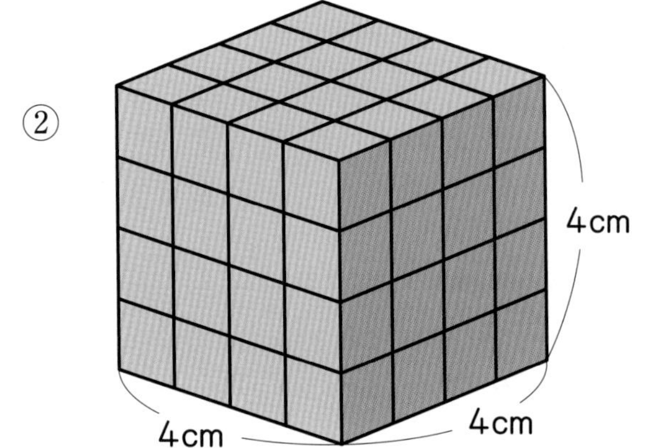

## Solids with the same volume

Let's create various solids by using 12 cubic blocks. Each cubic block is a 1 $cm^3$ cube.

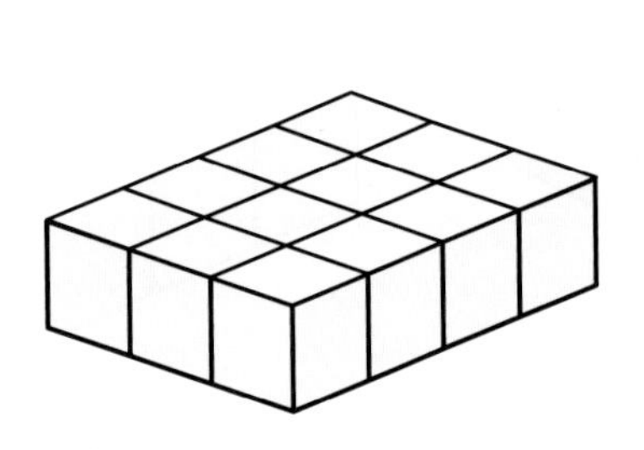

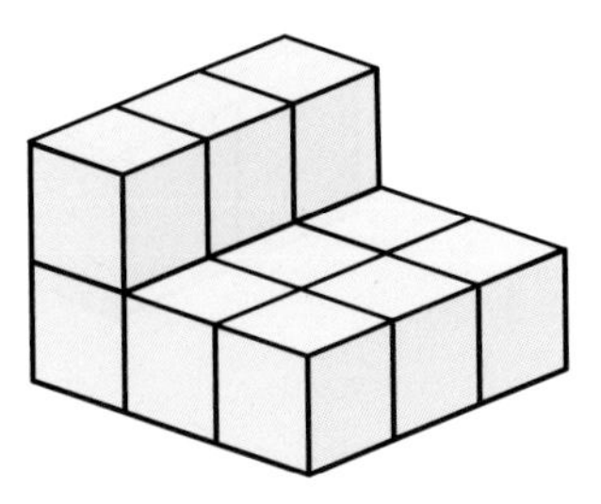

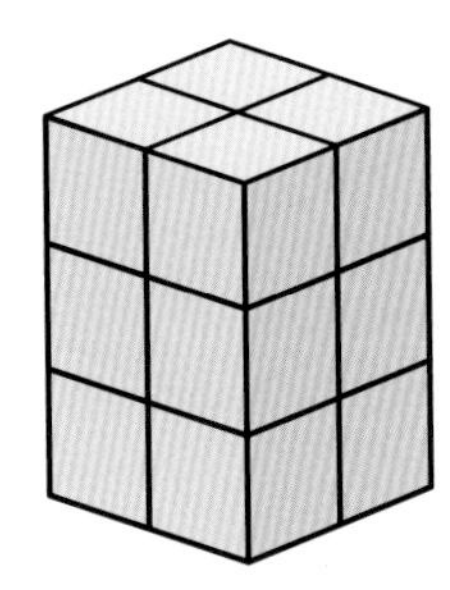

## 2 Formulas for volume

Want to know How to find the volume

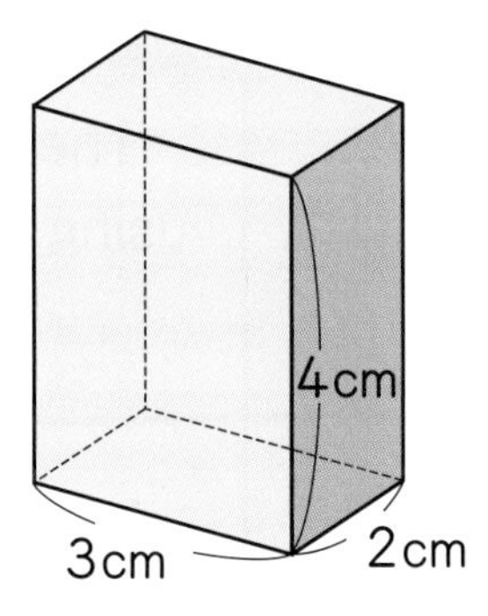


**1** **Let's think about how to find the volume of the cuboid shown on the right.**

Nanami

How many 1 $cm^3$ cubes are needed?

There is a formula to find the area, but...

Hiroto

**Purpose** How should we find the volume of a cuboid or cube?

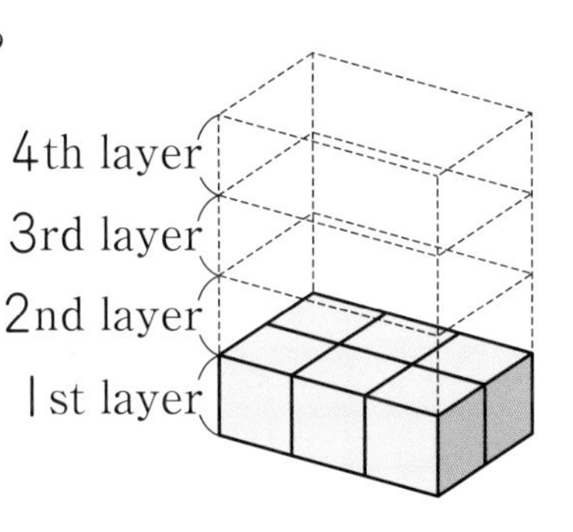


① How many 1 $cm^3$ cubes are there in the first layer?

② How many layers are there in total?

③ How many 1 $cm^3$ cubes are there in total?

$$2 \times 3 \times 4 = \square \text{ (cubes)}$$

Number for length, Number for width, Number for height, Total number

④ The number of cubes used for the length, width, and height is equal to the length, width, and height of the solid, respectively. Let's find the volume.

$$2 \times 3 \times 4 = \square \text{ (cm}^3\text{)}$$

Length, Width, Height, Volume

Want to extend

**1** Let's find the volume of the cube shown below in the same way as in **1**.

① How many 1 $cm^3$ cubes are there in the first layer?

② How many layers are there in total?

③ How many 1 $cm^3$ cubes are there in total?

Also, how many $cm^3$ is the volume?

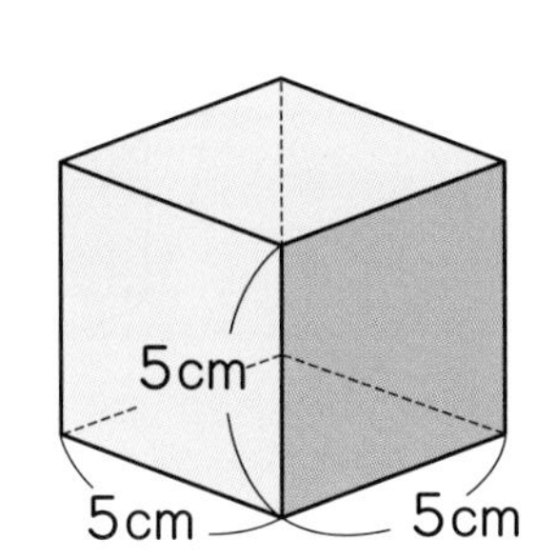

Summary

The volume of a cuboid is expressed by the following formula by using length, width, and height.

**Volume of a cuboid = length × width × height**

Since the length, width, and height of a cube are equal, its volume is expressed by the following formula.

**Volume of a cube = side × side × side**

Want to confirm

**2** Let's find the volume of the following cuboids and cube.

①

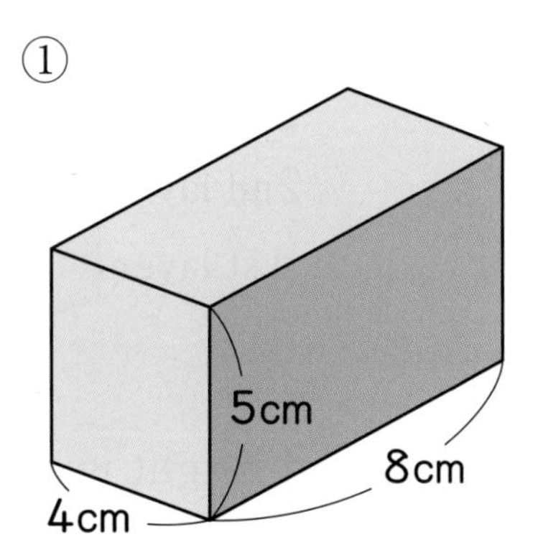


②

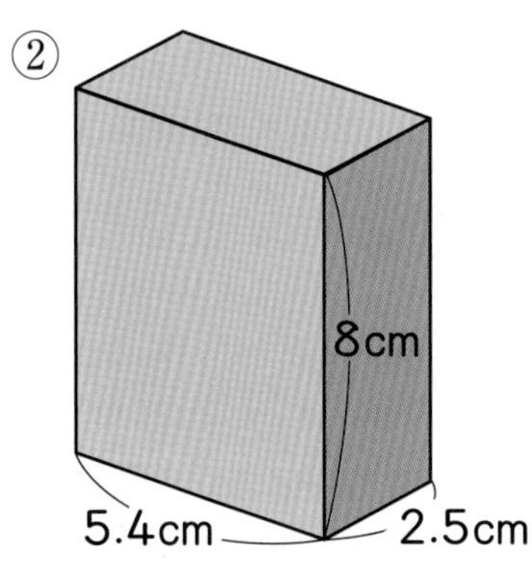


③

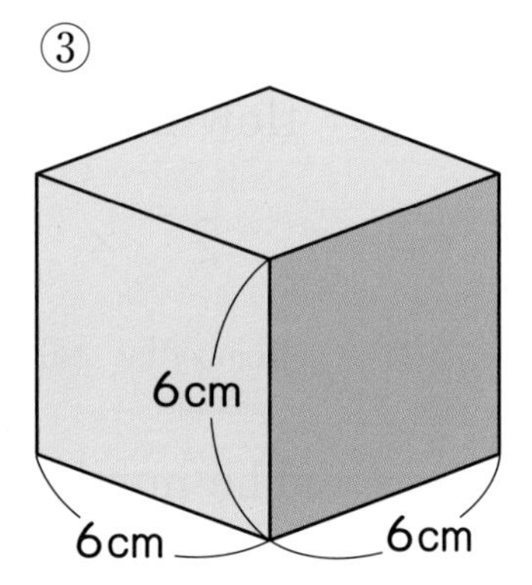


**3** Let's find the volume of the cuboid that can be assembled with the following net.

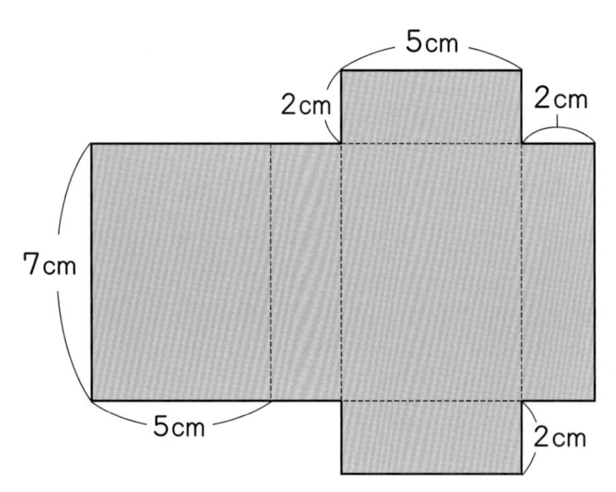


The number "2" and "3" in the unit of area "$cm^2$" and the unit for volume "$cm^3$" represent the multiplication of the length (cm) 2 and 3 times respectively.

Want to explore Relationship between height and volume

## 2

**The diagram on the right shows a cuboid with a length of 5 cm, width of 3 cm, and a height that is changing 1 cm, 2 cm, ... At this time, let's explore the relationship between each volume.**

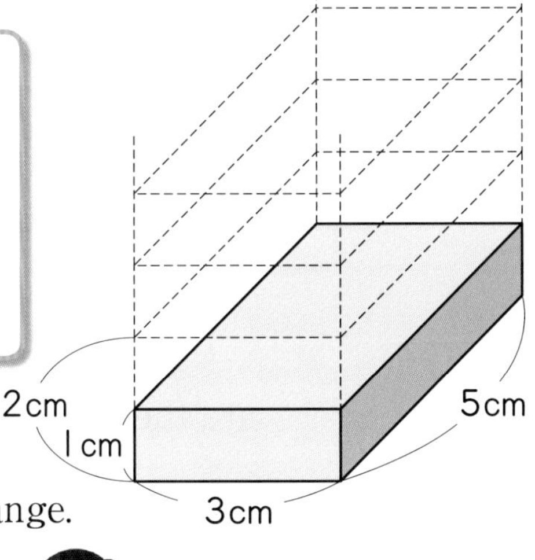


① Let's write the formula for the volume of a cuboid and explore the expressions that don't change.

Since, volume = length × width × height...

If only the height changes, what happens to the volume?

**Purpose** When the length and width of a cuboid don't change, what relationship is there between the height and volume?

② Let's consider the height as □ cm and the volume as ○ $cm^3$, and write a math sentence to find the volume.

③ Let's summarize the relationship between the height and volume of the cuboid in the following table.

Height and volume of the cuboid

| Height □(cm) | 1 | 2 | | | | | |
|---|---|---|---|---|---|---|---|
| Volume ○($cm^3$) | 15 | | | | | | |

Want to explain

④ From the table in ③, let's explain what relationship there is between the height and volume of the cuboid.

**Summary**

When the length and width of a cuboid don't change, the volume is proportional to the height.

Way to see and think 

In a cuboid, what happens to the volume if the length or width changes?

⑤ How many cm is the height of the cuboid when the volume is 150 $cm^3$? Let's think by using diagrams and tables.

## 3 Large volumes

Want to know

**1** **Let's think about how to represent the volume of the large cuboid shown on the right.**

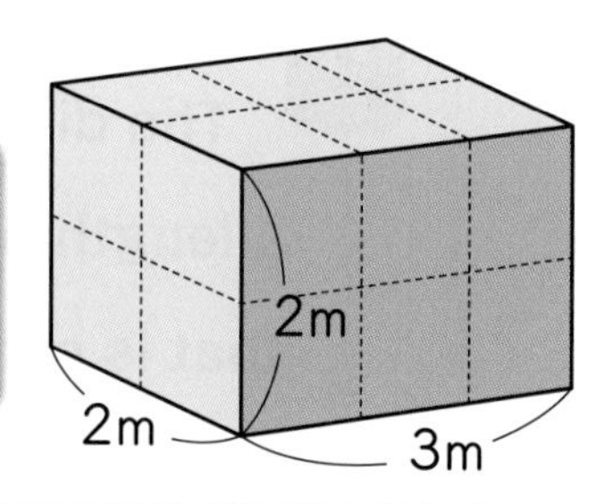


① How many cubes with a side of 1 m are there?

When the area was large, we represented it with "$m^2$."

The volume of a cube with sides of 1 m is called 1 **cubic meter** and is written as 1 $m^3$.

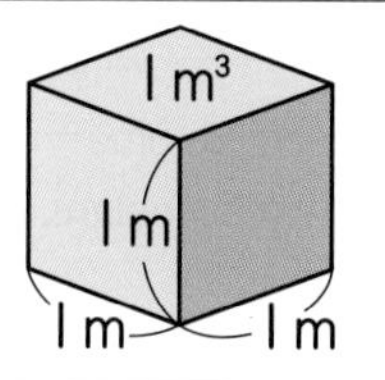


Way to see and think

It's represented in the same way as $cm^3$.

② How many $m^3$ is the volume of the above cuboid?

Want to explore

**1** Let's explore how many $cm^3$ is 1 $m^3$.

① How many 1 $cm^3$ cubes are placed along the length and width for the first layer?

② How many layers are there in total?

③ How many 1 $cm^3$ cubes are there in total? Also, how many $cm^3$ is the volume?

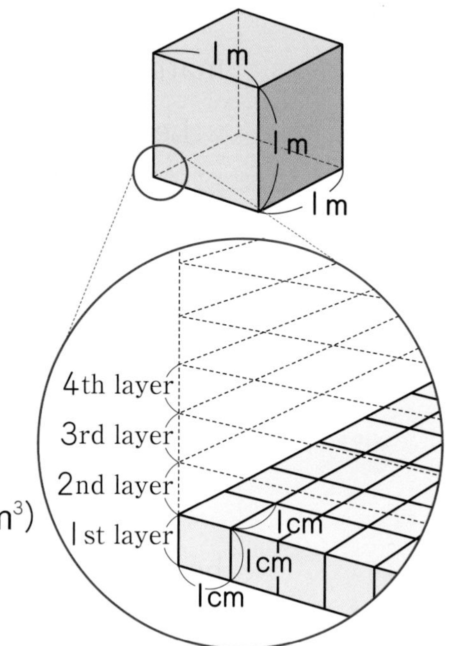


$100 \times 100 \times 100 = \square$ ($cm^3$)

Length Width Height Volume

**1 $m^3$ = 1000000 $cm^3$**

Want to think

Let's find the volume of the cuboid shown on the right.

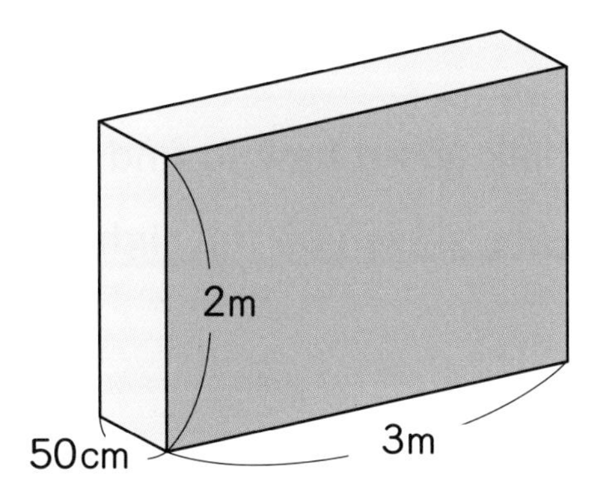


① Let's think about how to calculate.

② How many $m^3$ is the volume of the cuboid? Also, how many $cm^3$?

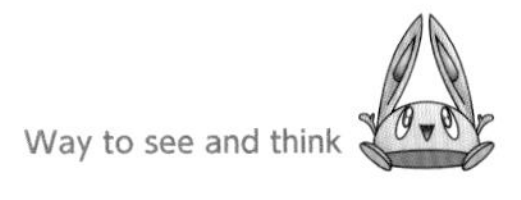


You can calculate by aligning to either cm or m.

Want to confirm

How many $m^3$ is the volume of cuboids ① and ②? Also, how many $cm^3$? Let's find each.

①

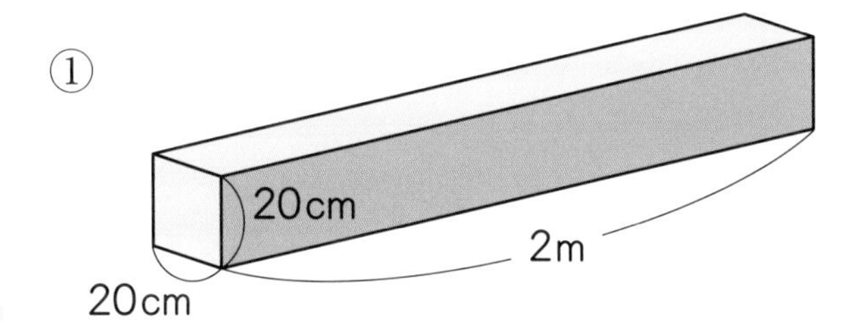


②

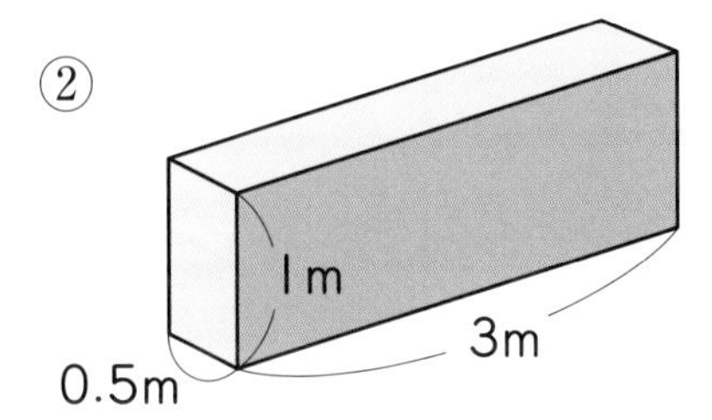


That's it

**Size of a $1m^3$ cube**

Let's predict how many people will fit in a $1$ $m^3$ cube. Let's try to actually confirm.

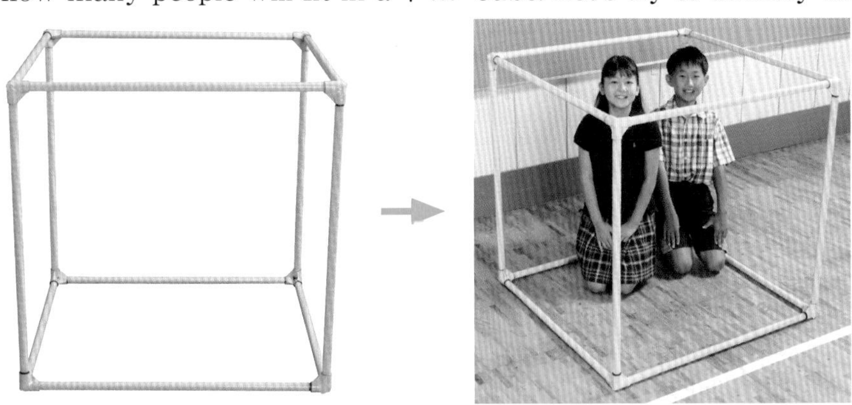

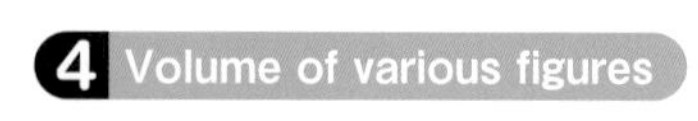


Want to know

Activity

## 1 Let's think about how to find the volume of the figure shown on the right.

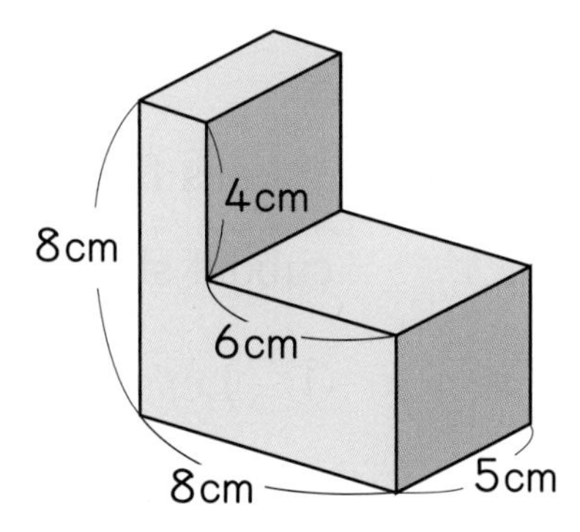

Nanami: In 4th grade, we found the area of this figure.

Hiroto: Since the volume of a cuboid can be found...

**Purpose** How should we find the volume of a figure that is not a cuboid nor cube?

① Write down your own idea in the diagram below.

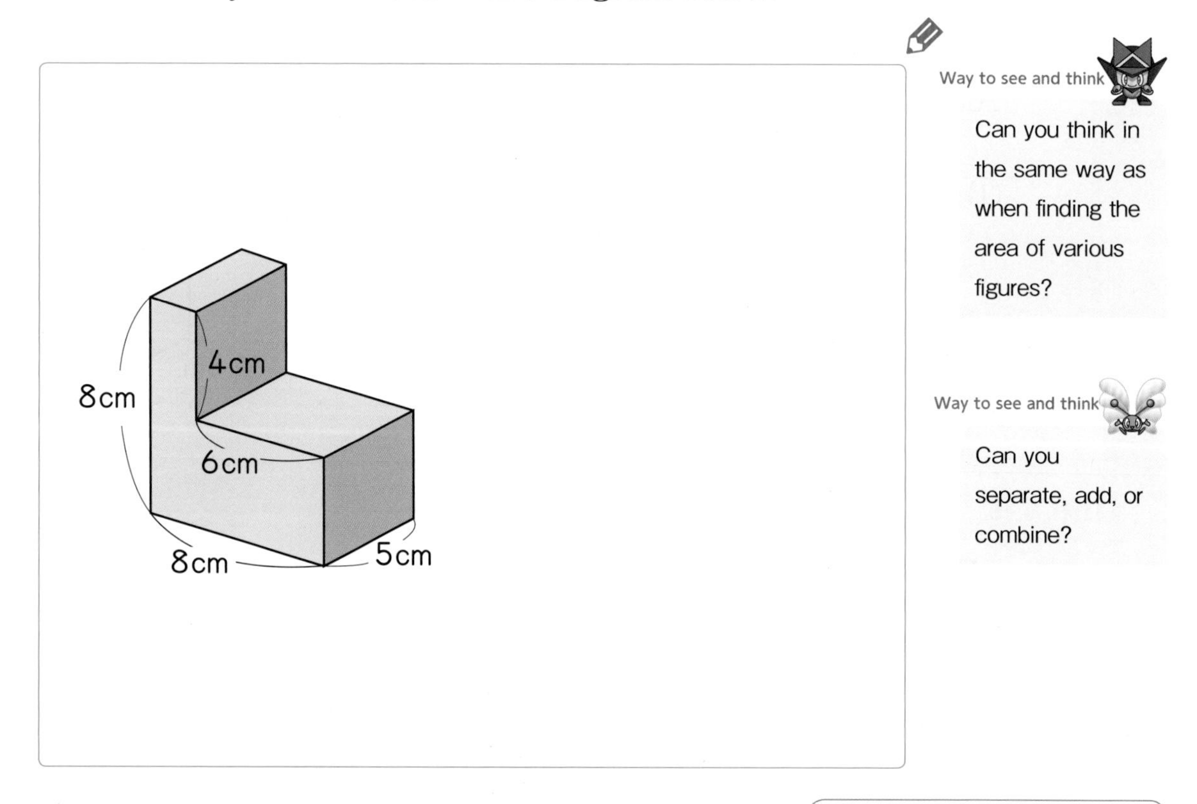


Way to see and think

Can you think in the same way as when finding the area of various figures?

Way to see and think

Can you separate, add, or combine?

Want to explain

② Let's explain the idea to your friends.

Daiki: If I separate the figure into two cuboids...

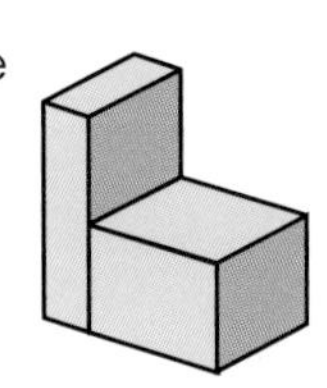

Want to compare

③ Let's compare the ideas of the following children. Also, let's write a math expression for each way of thinking and find the volume.

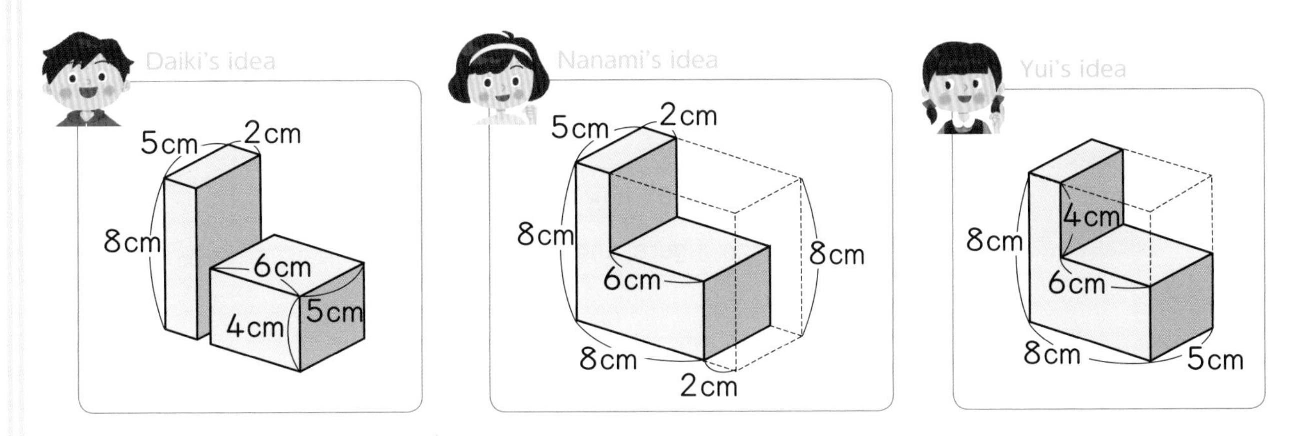


④ Let's discuss other methods with your friends.

Want to confirm

**1** Let's find the volume of the following figures.

①

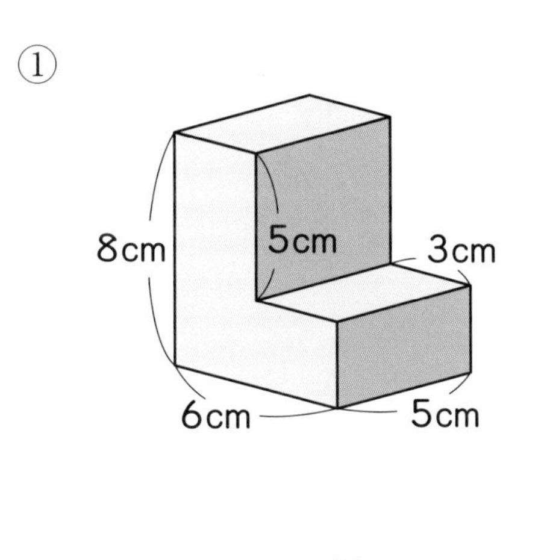


②

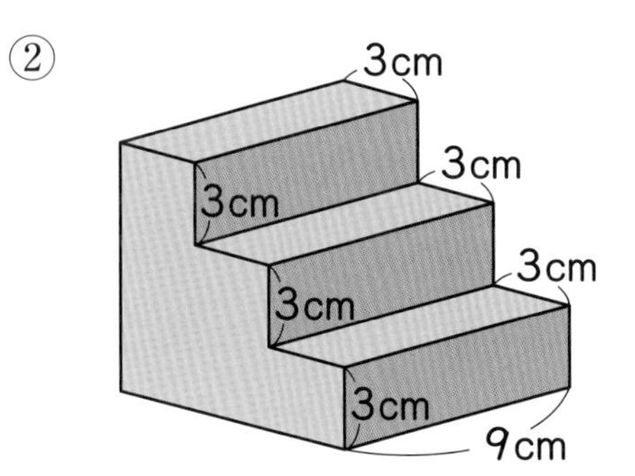


③

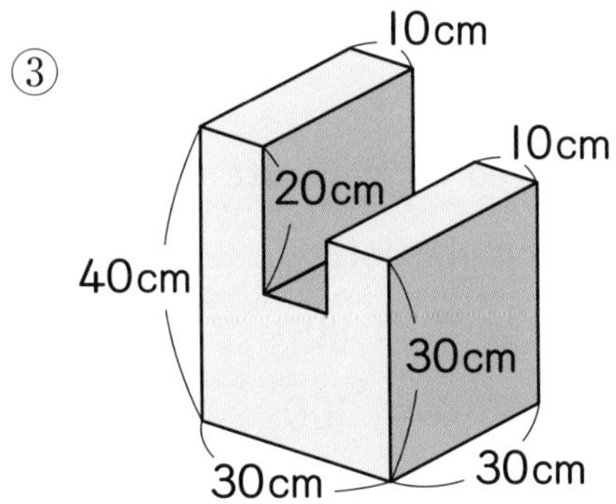


④

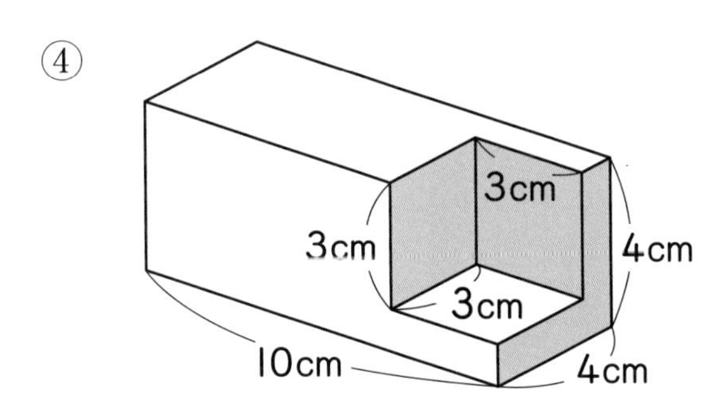

Become a writing master.

# Notebook for summarizing

## Let's summarize what you have learned on the day.

Write today's date.

January 25

8cm 4cm 5cm 7cm 5cm

Write the problem.

Let's think about how to find the volume of the figure shown on the right?

Let's learn based on the purpose.

Purpose: Let's use learned ideas to find the volume of the figure.

Organize and write your thoughts.

〈My idea〉

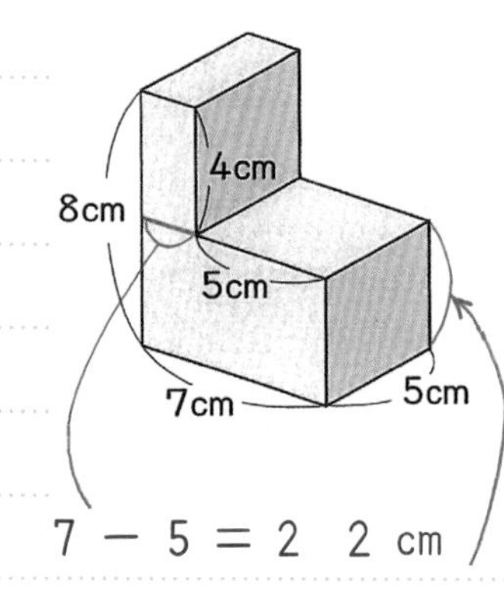


Separated the figure into two cuboids.

5 × 2 × 4 = 40

5 × 7 × ~~5~~ 4 = ~~175~~ 140

7 − 5 = 2　2 cm

If you make a mistake, do not erase it, just draw a line or write in red.

Here, I made a mistake with 5 cm.

8 − 4 = 4　4 cm

Must write why you made the mistake.

〈Takeshi's idea〉

8cm 4cm 5cm 7cm 5cm

Considered to separate as shown on the left.

5 × 2 × 8 = 80

5 × 5 × 4 = 100

80 + 100 = 180

Answer: 180 $cm^3$

〈Friends' idea〉

Yui's idea

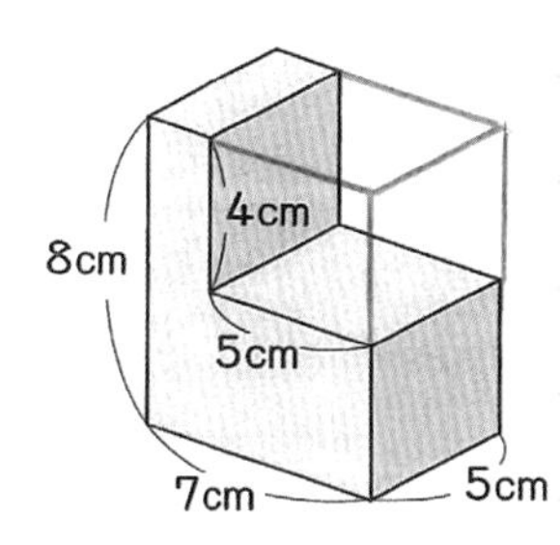


Subtract the small cuboid from the big cuboid.

$5 \times 8 \times 8 = 320$

$5 \times 6 \times 4 = 120$

$320 - 120 = 200$

Answer: 200 $cm^3$

Riku's idea

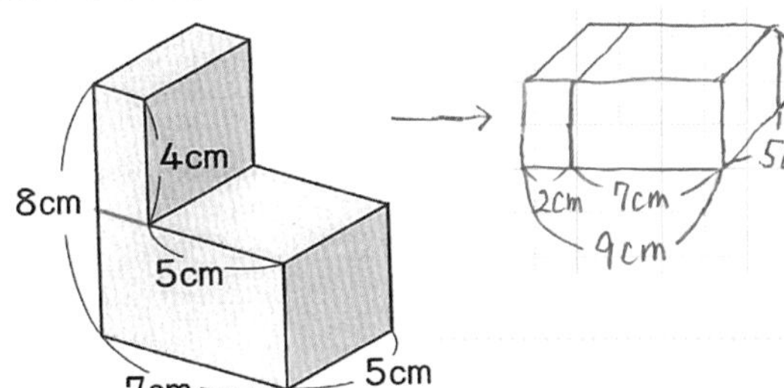


Rejoin the upper cuboid with the lower cuboid.

$5 \times (2 + 8) \times 4 = 200$

Answer: 200 $cm^3$

〈Reflect〉

○ The volume of the figure can be understood based on cuboids or cubes.

○ I noticed that Riku's idea is a method that cannot be used always. I think that is incredible.

〈Hinata's idea〉

Separated in the same way as I did but considered only one math sentence.

$5 \times 2 \times 4 + 5 \times 8 \times 4 = 200$

Answer: 200 $cm^3$

Write the classmate's ideas you consider good.

As for reflection, the following must be written.

- what you understood.
- what you noticed,
- what you are able to do,
- what you didn't understand,
- what you want to do more.

Want to know Volume and amount of water

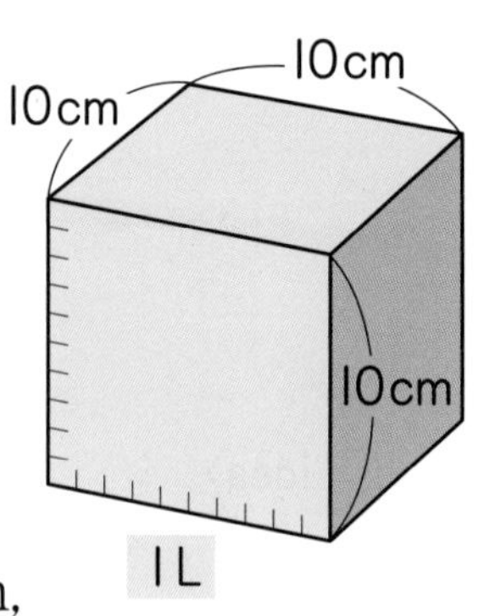


## 1 Let's explore the 1 L measuring container shown on the right.

① The 1 L measuring container is a cube with a length, width, and height of 10 cm. How many $cm^3$ is the volume of the 1 L measuring container?

1 L = ☐ $cm^3$

② 1 L = 1000 mL, how many $cm^3$ is 1 mL?

1 mL = ☐ $cm^3$

③ How many liters of water could fill a 1 $m^3$ tank?

1 $m^3$ = ☐ $cm^3$

1 $m^3$ = ☐ L

1000 L is also represented as 1 kL.

Way to see and think

The same as for length and weight, if a unit uses "k" it means 1000 times of the unit.

Units of amount of water are also expressed in kL, L, dL, and mL.

The relationship between the units of volume and amount of water is as follows.

**1000L = 1 $m^3$ 1 mL = 1 $cm^3$**

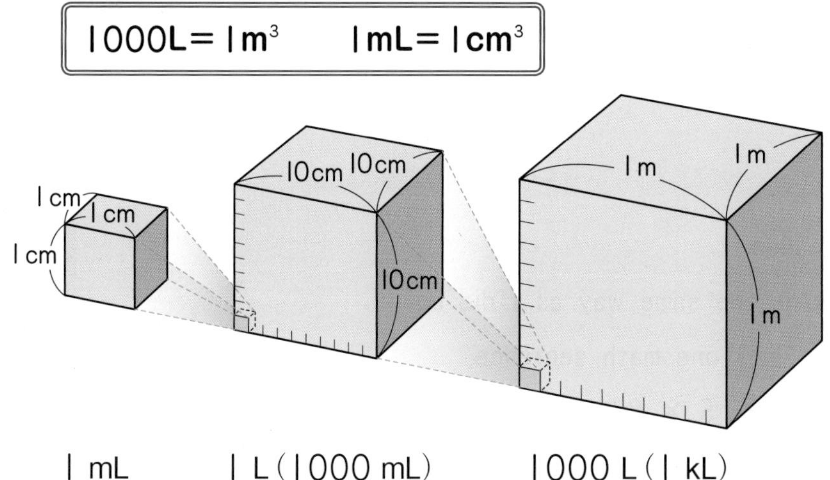


| 1 mL | 1 L (1000 mL) | 1000 L (1 kL) |
|---|---|---|
| 1 $cm^3$ | 1000 $cm^3$ | 1 $m^3$ (1000000 $cm^3$) |

Want to summarize Relationship between volume, area, and length

**1** Let's explore the relationship between units of length, area, and volume that we have learned until now. Also, let's try to discuss what kind of relationship it is.

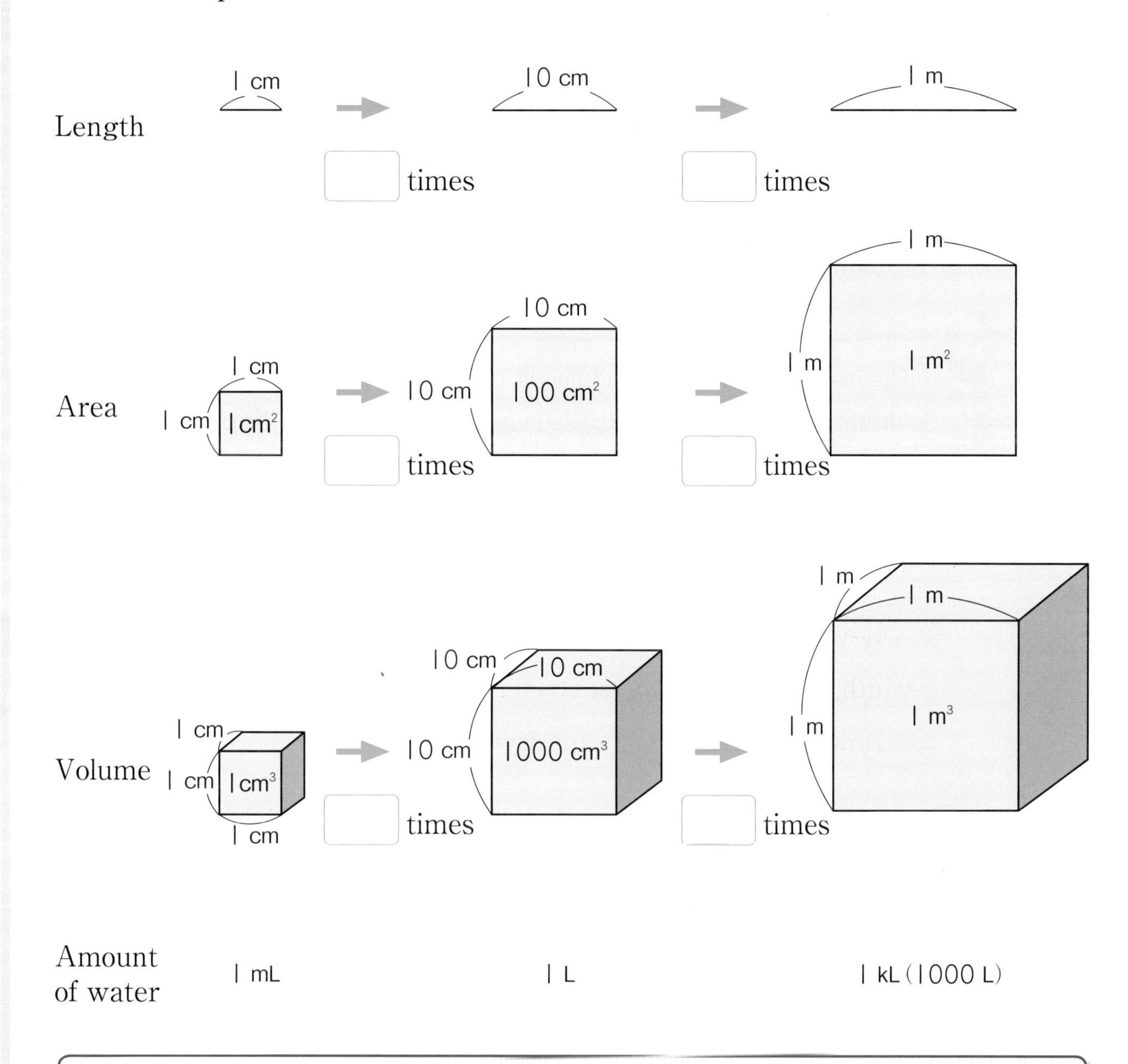


If you calculate 10 times of one side, the area becomes 100 times and the volume becomes 1000 times.

Want to know

**1** **There is a container with the shape of a cuboid that is made of 1 cm thick wood as shown on the right. How many $cm^3$ is the volume of water that fills this container?**

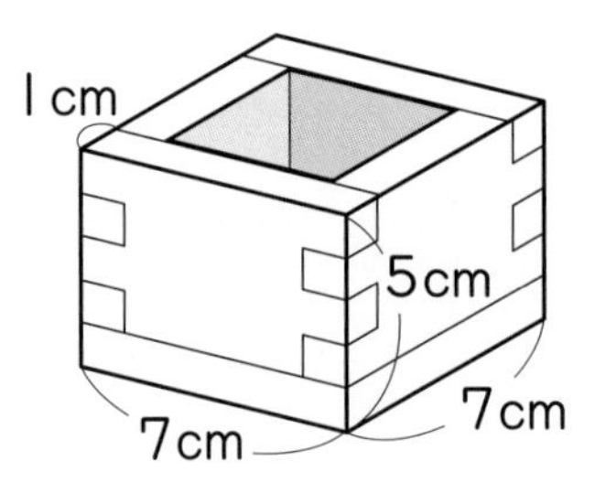


① What length should we know to calculate the volume of water that fills the container?

The inside length, width, and height of the container are called the **inside measures**. The inside height is also called the **depth**. The size of a container is measured with the volume of water that can fill it. This volume is the **capacity** of the container.

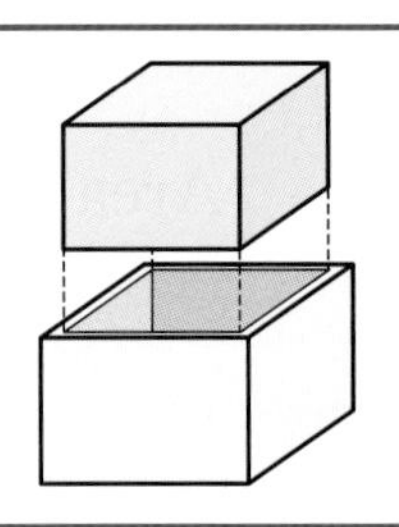

② How many cm are the inside length, inside width, and depth of the container?

③ How many $cm^3$ is the capacity of the container?

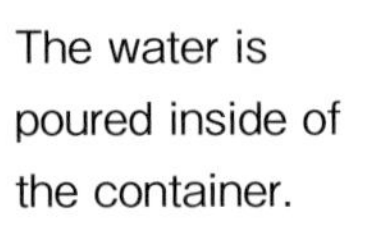

The water is poured inside of the container.

Yui

Want to confirm

**1** As shown in the diagram below, there is a water tank with the shape of a cuboid that is made of 1 cm thick glass. How many $cm^3$ is the capacity of the water tank?

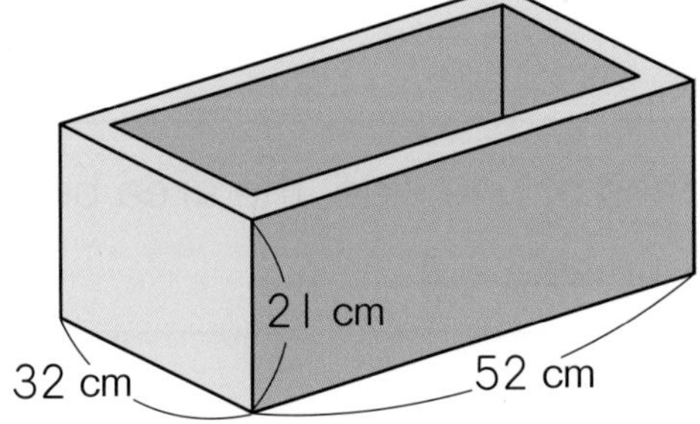

# What you can do now

**Can find the volume by using formulas.**

**1** Let's find the volume of the following cuboid and cube.

① 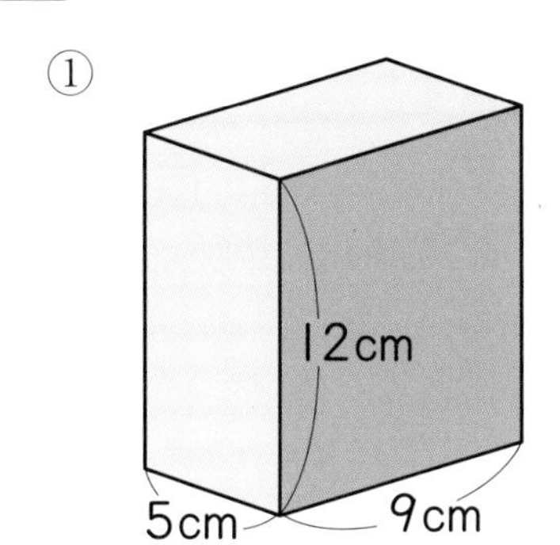


② 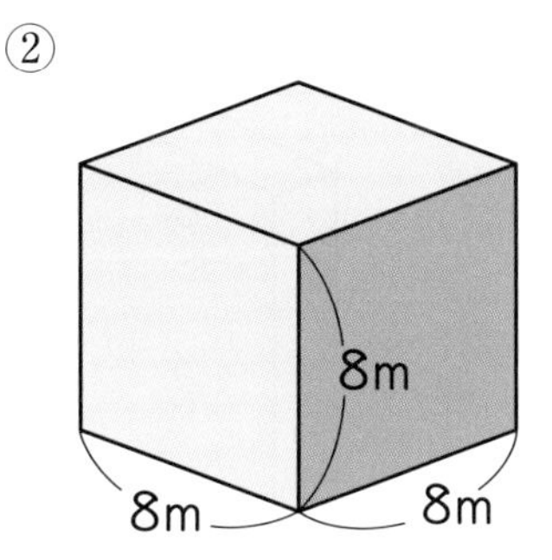


**Can find the volume by using learned ideas.**

**2** Let's find the volume of the following figures.

① 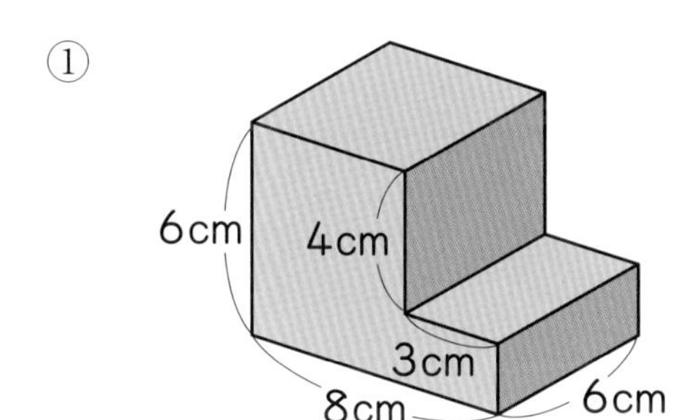


② 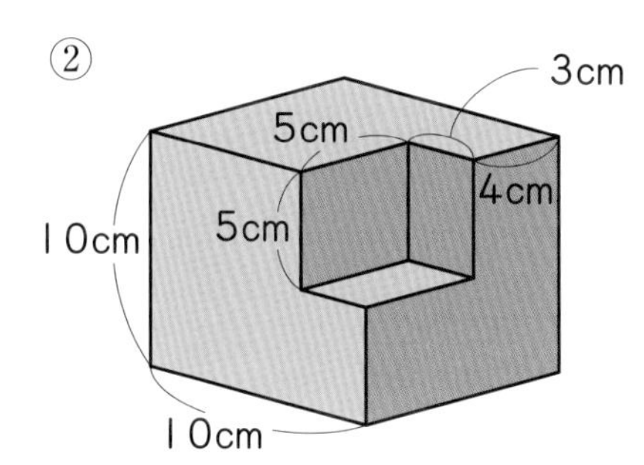


**Understanding the relationship between units of volume.**

**3** Let's write the numbers that apply in the following ☐.

① 1 L = ☐ mL ② 1 $m^3$ = ☐ $cm^3$

③ 1 kL = ☐ L ④ 1000 $cm^3$ = ☐ L

**Understanding about amount of water and capacity.**

**4** The figure shown on the right is a container with a length of 32 cm, width of 62 cm, height of 32 cm, and made with 1 cm thick board. Let's find the capacity of this container.

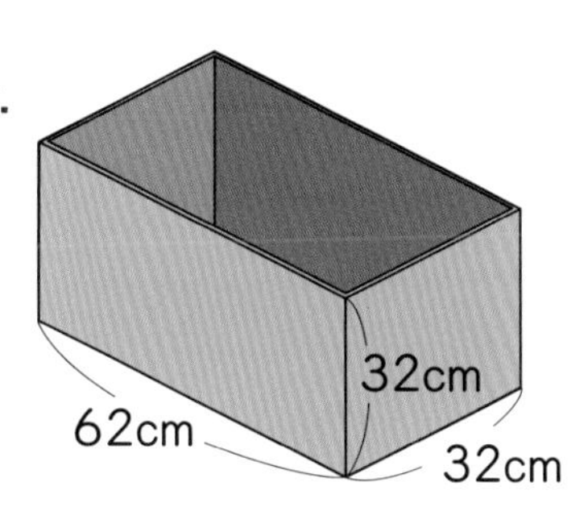


Supplementary Problems p.155

# Usefulness and efficiency of learning

**1** Let's find the volume of the following cuboids.

☐ Can find the volume by using formulas.

①

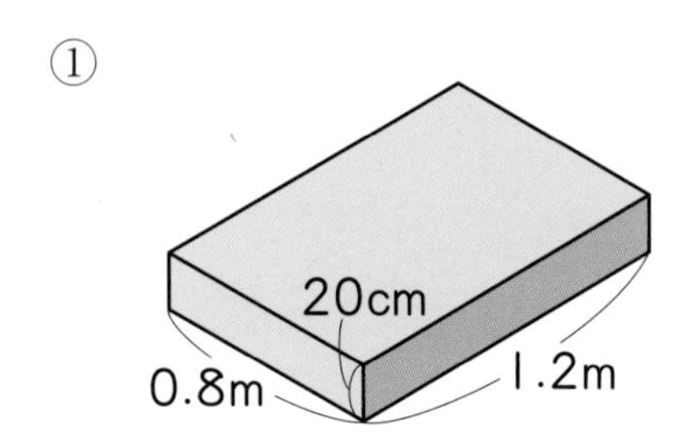


②

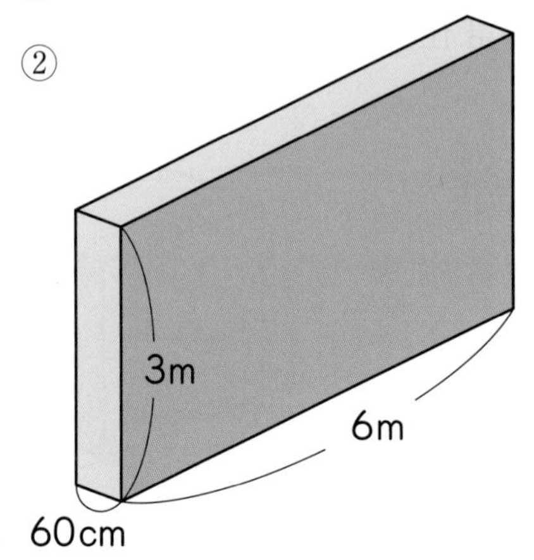


**2** Let's find the volume of the following figures.

☐ Can find the volume by using learned ideas.

①

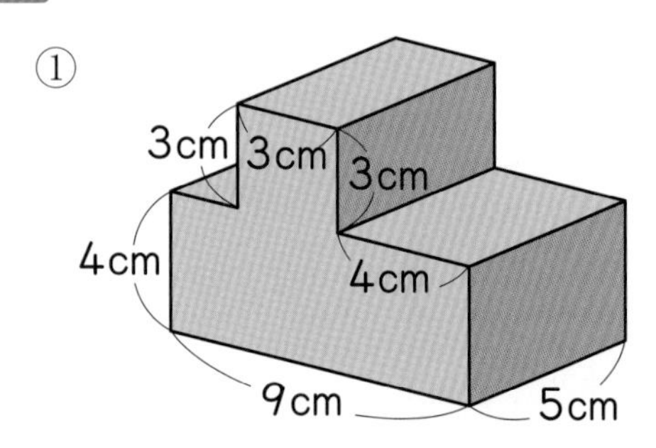


②

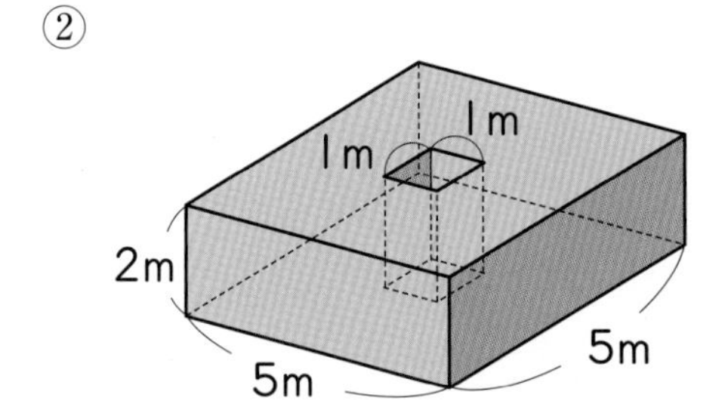


**3** How many $cm^3$ are 400 L of water? Also, how many $m^3$?

☐ Understanding the relationship between units of volume.

**4** Let's fill in the cuboid tank shown below with water.

How many times do we need to pour water with a 10 L bucket to fill completely the tank?

☐ Understanding about amount of water and capacity.

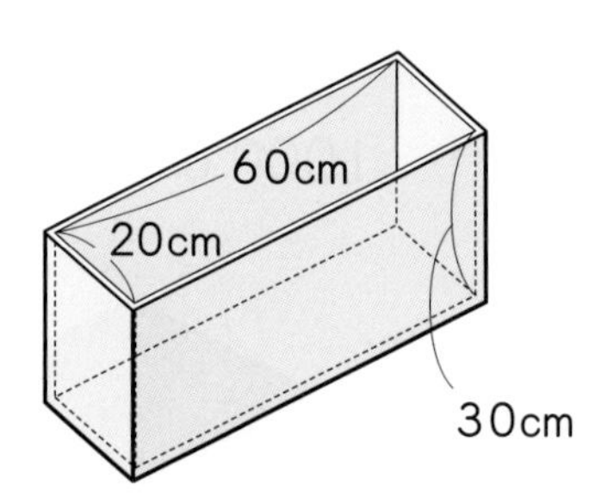

**Let's deepen.**

We can find the volume of objects enclosed by planes. What about irregular things?

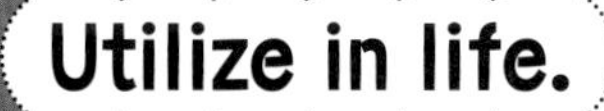


# Deepen.

## Let's measure volume.

**Want to find**

**Things always have a volume. How can we find the volume of objects that are not a cuboid or cube?**

**Let's think about how to find the volume of an irregular stone as the one shown on the right.**

Since it is irregular, it cannot be called a cuboid.

Since the volume can be considered as an amount of water...

① The following happened after immersing the stone in the water. Let's discuss what kind of things you noticed.

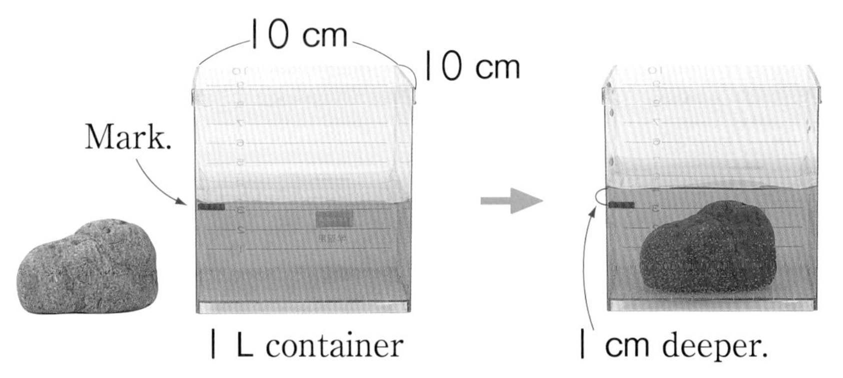


② When you place something in the water, the depth of the water increases by the volume. Let's find the volume of the stone in ①.

**Want to deepen**

**Can you find the volume of what other kind of things? Let's explore the volume of various things by using water.**

Reflect
Connect

**Problem**

## Will the capacity be the same?

The following container was made after cutting 4 squares from a rectangular board with a length of 12 cm and a width of 20 cm.

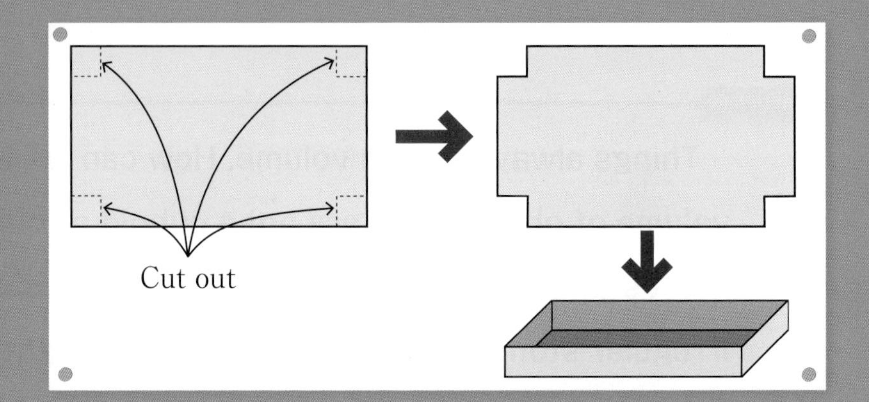
Cut out

◎ If we change the length of the side of the squares to be cut out, will the capacity of the container change?

Let's explore the capacity of the containers when the length of the side of the squares to be cut out are 1 cm and 2 cm.

●Let's confirm the inside length, inside width, and depth of each container.

When the side of the square is 1 cm

inside length ☐ cm
inside width ☐ cm
depth ☐ cm

When the side of the square is 2 cm

inside length ☐ cm
inside width ☐ cm
depth ☐ cm

We can find the capacity with "length × width × depth."
Hiroto

The depth has the same length as the side of the square.
Daiki

If you change the side of the square, then the inside length, inside width and depth will change.
Nanami

●Let's find the capacity of each container.

When the side of the square is 1 cm

Math expression: ＿＿＿＿ Answer: ＿＿＿＿

When the side of the square is 2 cm

Math expression: ＿＿＿＿ Answer: ＿＿＿＿

Containers made from the same rectangular board have different capacities.

> ◎ If we increase the length of the side of the squares to be cut out, will the capacity of the container also increase?

●Let's explore the case when the length of the side of the squares to be cut out is 3 cm, 4 cm, and 5 cm.

When the side of the square is 3 cm

Math expression: ＿＿＿＿ Answer: ＿＿＿＿

When the side of the square is 4 cm

Math expression: ＿＿＿＿ Answer: ＿＿＿＿

When the side of the square is 5 cm

Math expression: ＿＿＿＿ Answer: ＿＿＿＿

> ◎ Even if the original figure was a square board with a side of 15 cm, will the capacity change?

Depending on the length of the side of the squares to be cut out, you can make either a cuboid or a cube container.

「Volume of cube = side × side × side」

Want to connect

Yui

The capacity is the largest when the side of the squares is 2 cm. The capacity becomes smaller when the length of the side increases.

Yui

Even when the original figure is a square board, the capacity changes in the same way?

# How many people ride the train?

Problem Can we find the number of people by using ratios?

Ratio (2)

# 17 Let's think about how to compare two quantities on problems using ratios.

## 1 Ratio of two quantities

Want to explore

**1** **In Sakura's class, 16 children are boys and 20 children are girls. Let's explore the relationship between the number of boys and number of girls.**

Purpose How can we represent the relationship between two quantities?

① Let's find the ratio of the number of boys based on the number of girls.

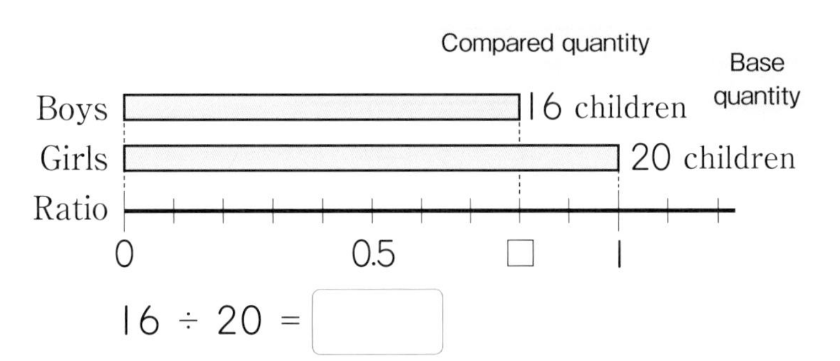


$16 \div 20 = \square$

| Base quantity | Compared quantity |
|---|---|
| 20 children | 16 children |
| 1 | □ |

Ratio

② Let's find the ratio of the number of girls based on the number of boys.

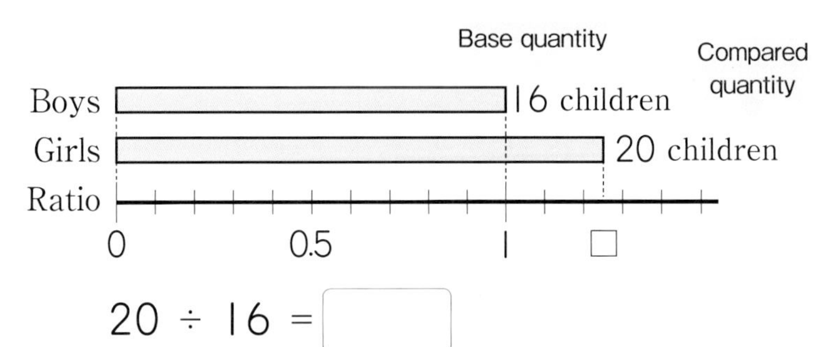


$20 \div 16 = \square$

| Base quantity | Compared quantity |
|---|---|
| 16 children | 20 children |
| 1 | □ |

Ratio

Summary

In addition to represent the relationship between total quantity and a part of the total, the relationship between two quantities can also be represented as a ratio. The ratio changes if the base quantity is changed. In some cases, the ratio becomes larger than 1.

Akira and Yuta are in the mini basketball club. The table on the right shows the number of shots that were scored by both in 3 games played in 4th grade and 5th grade.

From 4th grade into 5th grade, can you say which of the two players improved more the performance?

**Number of scored shots**

| | 4th grade | 5th grade |
|---|---|---|
| Akira | 20 | 50 |
| Yuta | 30 | 69 |

① Let's find the ratio of the number of scored shots in 5th grade based on the number of scored shots in 4th grade for each player.

Akira

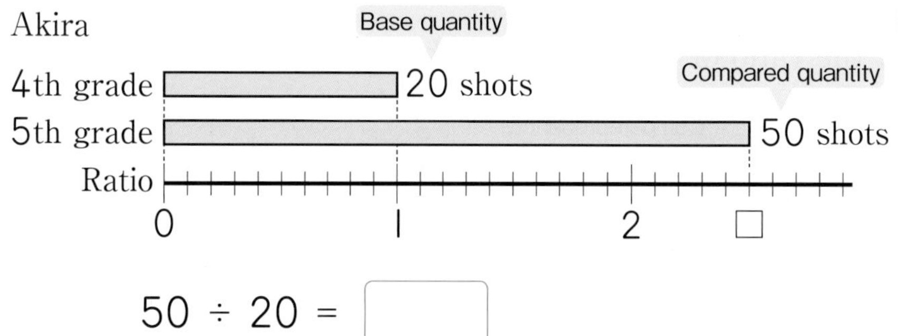


| Base quantity | Compared quantity |
|---|---|
| 20 shots | 50 shots |
| 1 | □ |

Ratio

$50 \div 20 = \square$

Yuta

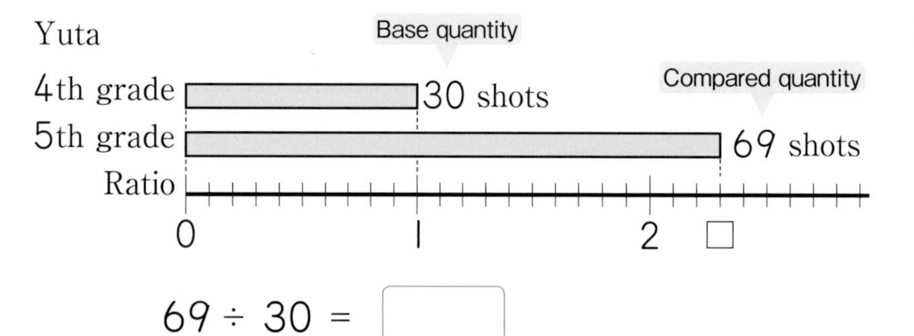


| Base quantity | Compared quantity |
|---|---|
| 30 shots | 69 shots |
| 1 | □ |

Ratio

$69 \div 30 = \square$

② Let's discuss which of the players improved more the performance.

Hiroto

What happens if I compare the difference between 4th grade and 5th grade?

We should think how many times of the result in 4th grade is the result in 5th grade.

Nanami

Way to see and think

You can think in the same way as with "times of" or "ratio" learned in 4th grade.

 Problems using ratios

Want to solve  Problems on finding compared quantities

**1** **A painter is painting a wall that has an area of 24 m². Until now, he has painted 25% of the wall. How many m² has he painted?**

① Let's change 25 % into a decimal number.

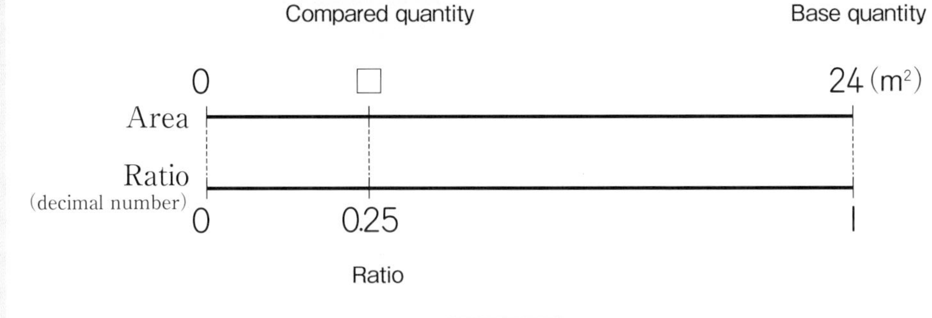


| Base quantity | Compared quantity |
|---|---|
| 24 m² | □m² |
| 1 | 0.25 |

Ratio

$24 \times 0.25 =$ □

② Let's find the area of 1 % of the wall.

1 % of the area is,

$24 \div 100 =$ □

25 % of the area is,

□ $\times 25 =$ □

| | Compared quantity |
|---|---|
| 0.24 m² | □m² |
| 1 % | 25 % |

Ratio

You can find the compared quantity with the following math sentence.

Compared quantity = base quantity × ratio

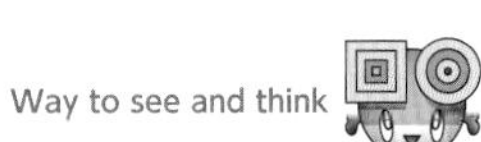


When you reverse the math sentence to find the ratio, it is changed into the math sentence to find the compared quantity.

Want to confirm

 There is a train with a capacity of 80 passengers in each car. When the crowdedness of a car is 110 %, how many passengers are there in the car?

Want to solve

Activity

## 2

**Yuki's mother bought a shirt with a 20% discount that had an original price of 1500 yen. How many yen was the discounted price?**

① How many yen was the discount?

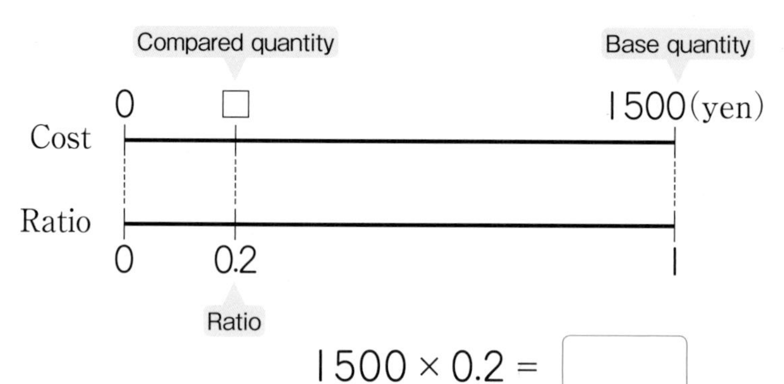


Base quantity | Compared quantity

| 1500 yen | □ yen |
|---|---|
| 1 | 0.2 |

Ratio

$1500 \times 0.2 = \square$

Base quantity Ratio Compared quantity

Want to explain

② The following children thought about finding the cost of the shirt. Let's explain the ideas of the children and find the cost.

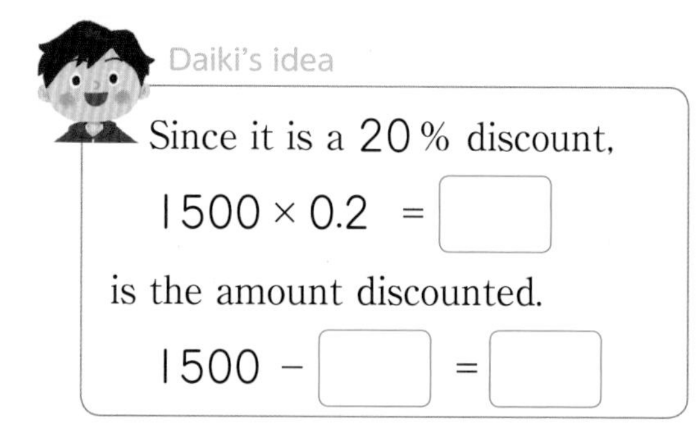

Daiki's idea
Since it is a 20 % discount,
1500 × 0.2 = □
is the amount discounted.
1500 − □ = □


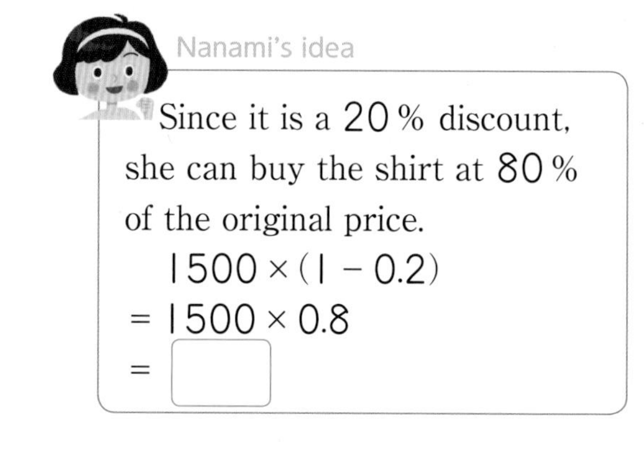

Nanami's idea
Since it is a 20 % discount,
she can buy the shirt at 80 %
of the original price.
1500 × (1 − 0.2)
= 1500 × 0.8
= □


Want to confirm

Usually, a 150 g snack is sold with an increased weight of 20 %. How many grams is the weight?

Want to try

We bought some goods for 400 yen and would like to sell them with an added 25 % profit. How many yen should the selling price be?

## 3

**At Masaki's house, a part of a field has been turned into a flower garden. The area of the garden is 60m$^2$, which is 20% of the total area of the field. How many m$^2$ is the area of the field?**

① Let's change 20% into a decimal number.

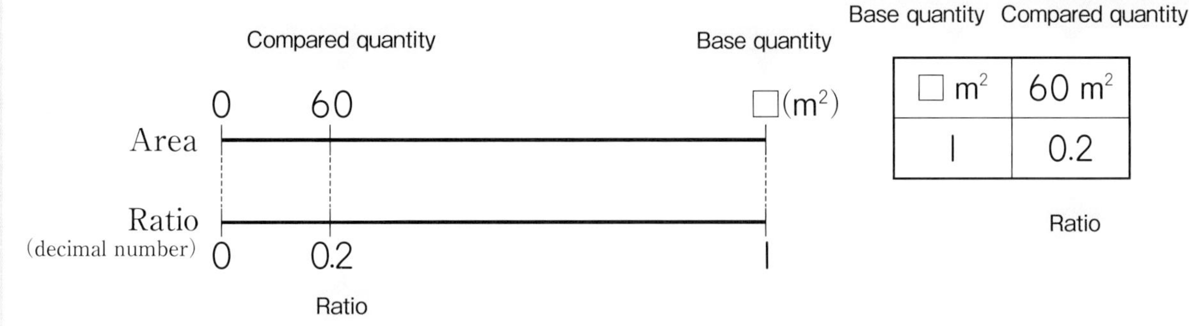


| □ m$^2$ | 60 m$^2$ |
|---|---|
| 1 | 0.2 |

If you consider the area of the field as □ m$^2$ and write a math sentence to find the compared quantity, $\square \times 0.2 = 60$

So, the math sentence to find □ is $60 \div 0.2 =$ [ ]

② Let's find the area of 1% of the field.

1% of the field is $60 \div 20 =$ [ ]

100% of the field is [ ] $\times 100 =$ [ ]

Compared quantity

| □ m$^2$ | 60 m$^2$ |
|---|---|
| 1% | 20% |

Ratio

You can find the base quantity with the following math sentence.
Base quantity = compared quantity ÷ ratio

Want to confirm

4 On a particular day, car number three on the bullet train had 102 passengers. This number of passengers is 120% of the capacity of the car. How many people is the capacity of car number three?

5 There is a lottery where 15% of the tickets are winning tickets. If there are 30 winning tickets, how many lottery tickets are there in total?

Want to solve Utilization of ratios

**4** **The North and South Supermarkets have the same bakery. On Sundays, the North Supermarket bakery has a 2割 (2 wari) discount on any bread. On Sundays, the South Supermarket bakery discounts 100 yen when the total cost is higher than 400 yen. From the following cases Ⓐ~Ⓓ, let's try to think which bakery has the cheapest way to buy.**

Ⓐ The case of buying three 150 yen bread.

Ⓑ The case of buying three 180 yen bread.

Ⓒ The case of buying three 200 yen bread.

Ⓓ The case of buying one of each: 150 yen bread, 180 yen bread, and 200 yen bread.

Want to find in our life

**6** From your surroundings, let's search for things that use ratios.

100% fruit juice

2割 discount

1割 discount

# What you can do now

☐ **Can compare two quantities by using ratios.**

**1** Shota has a 15 m tape. Rika has a 12 m tape. Let's answer the following questions.

① Let's find the ratio of the length of Rika's tape based on the length of Shota's tape.

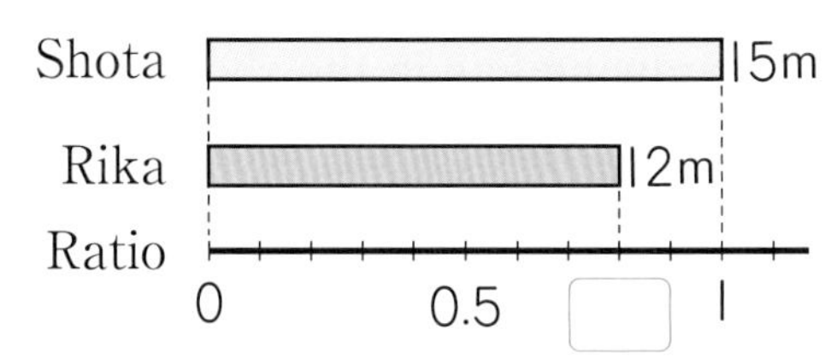


② Let's find the ratio of the length of Shota's tape based on the length of Rika's tape.

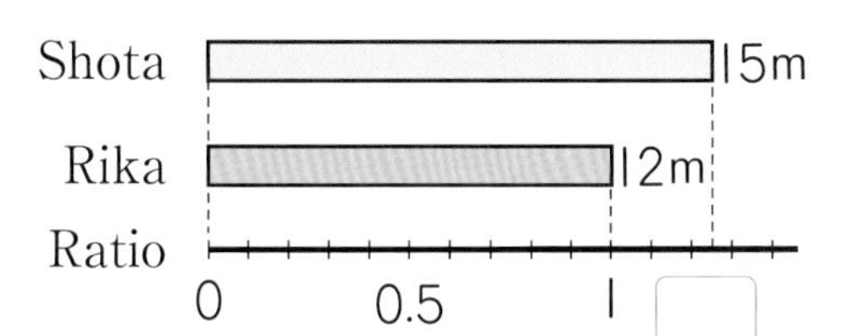


☐ **Can find ratios.**

**2** Last year, the number of 5th graders in Takuma's elementary school was 125 children. This year, the number increased by 10 children. Let's answer the following questions.

① How many children is the number of 5th graders this year?

② This year, 54 children in 5th grade are girls.
From the total 5th graders, what percentage is the ratio of girls?

③ What is the percentage of the number of children this year compared to the number of last year?

☐ **Can find the compared quantity.**

**3** I'm carrying 300 eggs. In the case that 4 % of the eggs were broken in the transportation process, can you think how many eggs were broken?

☐ **Can find the base quantity.**

**4** At a store, bags are discounted by 20 % and sold for 9800 yen. How many yen was the original price?

Supplementary Problems p.156

# Usefulness and efficiency of learning

**1** A 3000 yen canned food set is sold for 2100 yen. What was the percentage discounted from the original price?

☐ Can find ratios.

**2** You will be drawing lots in the shopping district of the town. 150 lots were prepared. Let's answer the following questions.

① You take 12 lots in the first turn. Let's represent the ratio of the first turn based on the total number of lots as a percentage.

② You take 14 % of the whole lot in the second turn and 30 % of the whole lot in the third turn. How many lots did you draw in the second and third turn?

☐ Can find the compared quantity.

**3** The snack's weight was increased by 35 % for a limited time. It became 189 g. How many grams is the normal weight?

☐ Can find the base quantity.

**4** In Natsuki's house, the area that is planted with tomatoes is 96 $m^2$. This is 30 % of the total field. How many $m^2$ is the area of the total field?

☐ Can find the base quantity.

**5** A colored pencil has an original price of 400 yen. This pencil is sold by the East Store with a discount of 80 yen and is sold by the West Store with a discount of 12 %. At which store is the pencil cheaper and by how many yen?

☐ Can find the compared quantity.

## Let's deepen.

I see a lot of "○ yen discount" and "△% discount." Actually, how much is the difference?

# What do you want to be?

Hiroto, what do you want to be in the future?
1

I like science experiments, so I want to be a researcher.
Also, I think that astronauts are cool.
2

I want to be an athlete and play abroad.
What about other people?
3

When I explored on the internet, I found a summary of "professions that elementary school students want to become in the future."
4

How should we look at the diagram we explored?

# 18 Various graphs

# Let's explore how to represent graphs using ratios.

## 1 Circle graph

Want to know

**1** **The graph on the right represents the ratio of sports where Japanese athletes have won gold medals at the Summer Olympic Games. Let's think about this graph.**

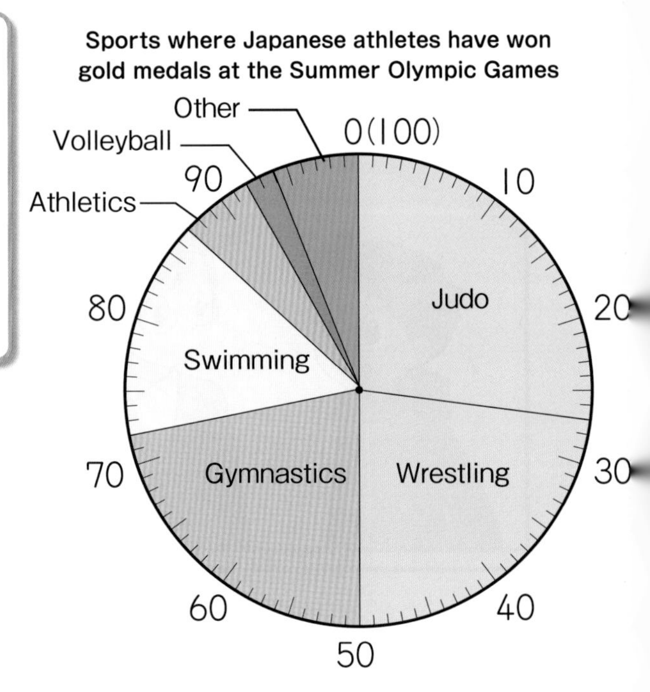


Purpose What kind of representation is a graph using circles?

A graph that is drawn as a circle is called **circle graph**.

The circle graph shows the ratio of each part to the whole, using the radius as divider.

① What percentage of the total is the ratio of gold medals won in judo?

② What percentage of the total is the ratio of gold medals won in each of wrestling, gymnastics, swimming, athletics, and volleyball?

③ What percentage of the total is other sports? Also, find the answer as a fraction.

Summary

In circle graphs, draw the ratios in descending order, and draw "other" at the end.

Also, when the order is significant, it is represented in that order.

Want to know

## 2

**The circle graphs below show the results of exploring the professions that elementary school students wanted to become in 2007 and 2017. Let's answer the following questions about these circle graphs.**

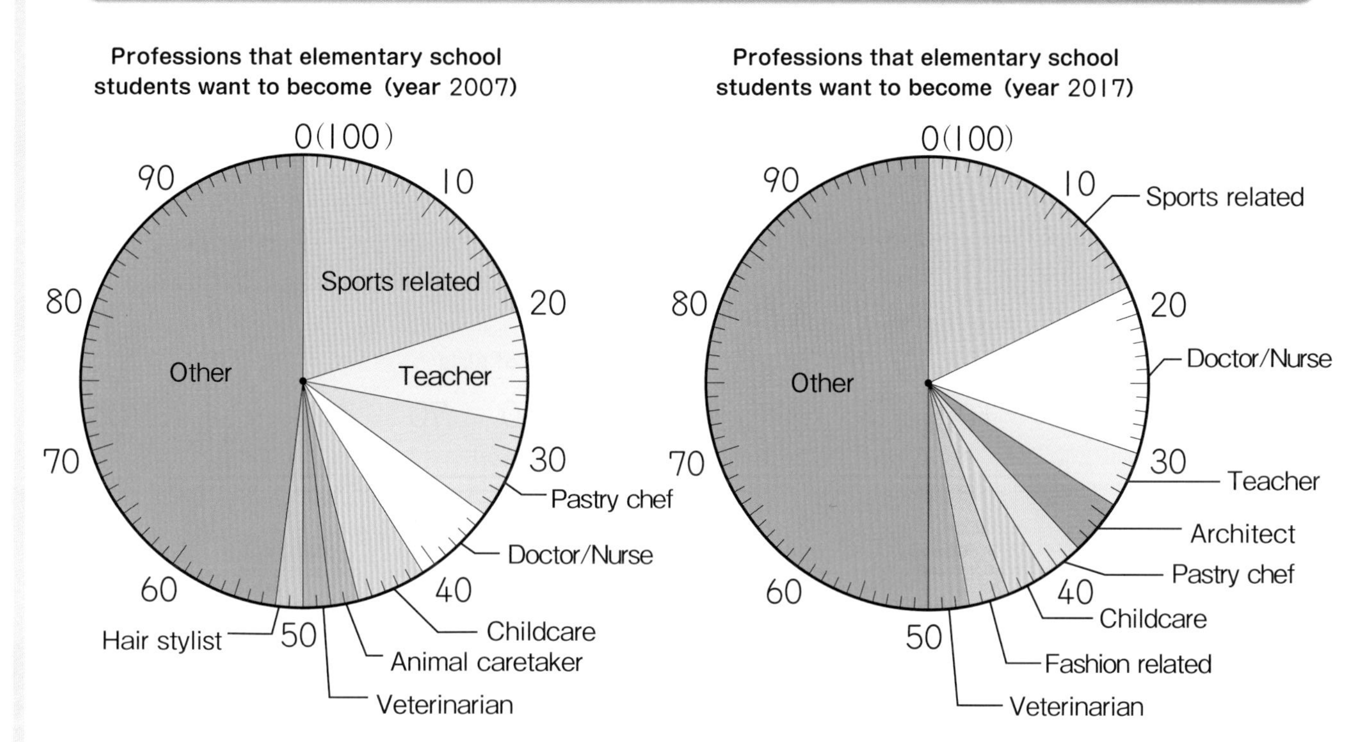


① Let's summarize in the table the ratio of professions that students wanted to become in 2007 and 2017.

**Professions that elementary school students want to become (year 2007)**

| Sports related | Teacher | Pastry chef | Doctor/Nurse | Childcare | Animal caretaker | Veterinarian | Hair stylist | Other |
|---|---|---|---|---|---|---|---|---|
| | | | | | | | | |

**Professions that elementary school students want to become (year 2017)**

| Sports related | Doctor/Nurse | Teacher | Architect | Pastry chef | Childcare | Fashion related | Veterinarian | Other |
|---|---|---|---|---|---|---|---|---|
| | | | | | | | | |

② How many times of the ratio of Doctor/Nurse chosen in 2007 is the ratio of Doctor/Nurse chosen in 2017?

③ For each year, what fraction of the total was chosen as others?

④ Let's try to discuss what you understood in ③.

Want to know

**1** **The following graphs represent the favorite Olympic events of 120 boys and 80 girls from 5th grade. Let's think about these graphs.**

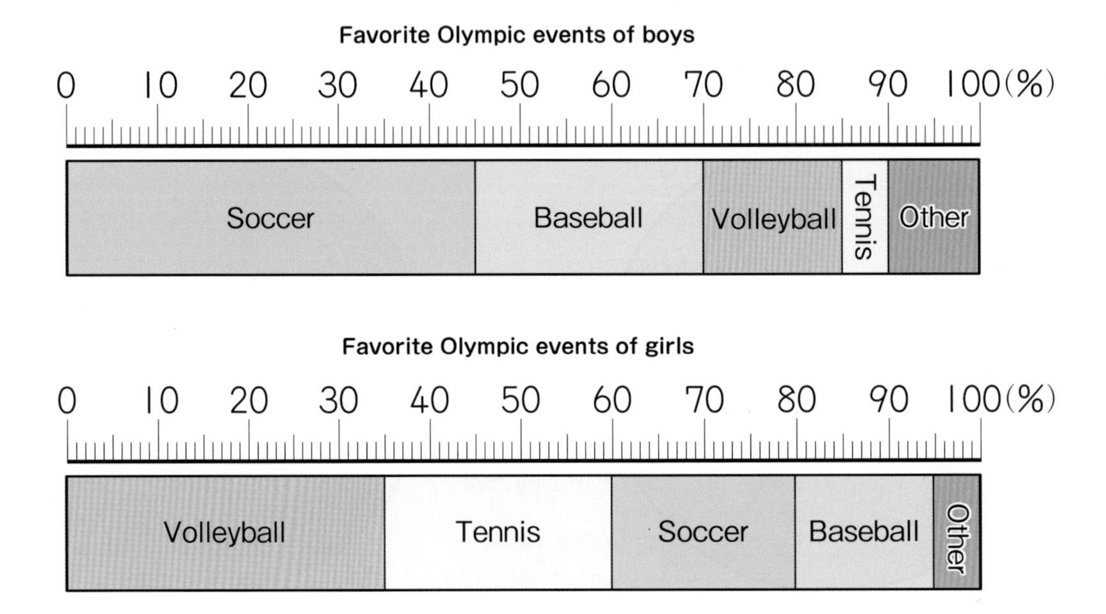


It uses ratios in the same way as circle graphs, but the shape is different.

A graph that represents the part-whole relationship by rectangle-like bands is called **strip graph**.

With a strip graph, it is easy to see the ratio of each part to the whole because the size of each part is shown by the corresponding area of the rectangle.

Want to think

① What percentage of the total is the ratio of volleyball chosen by each boys and girls?

② What percentage of the total is other sports in each graph? Also, find the answer as a fraction.

③ Let's summarize the number of people that chose each event. Also, let's find the total number of boys and girls for each event.

**Favorite Olympic events**

| | Soccer | Baseball | Volleyball | Tennis | Other | Total |
|---|---|---|---|---|---|---|
| Boys (people) | | | | | | |
| Girls (people) | | | | | | |
| Total (people) | | | | | | |
| Total (%) | | | | | | |

④ Let's find the percentage for each event based on the total 5th graders.

Want to represent

⑤ The girls' graph was re-drawn based on the order of events from the boys' graph. Let's represent the total results in the same order of events from the boys' graph and draw dotted lines to connect each event.

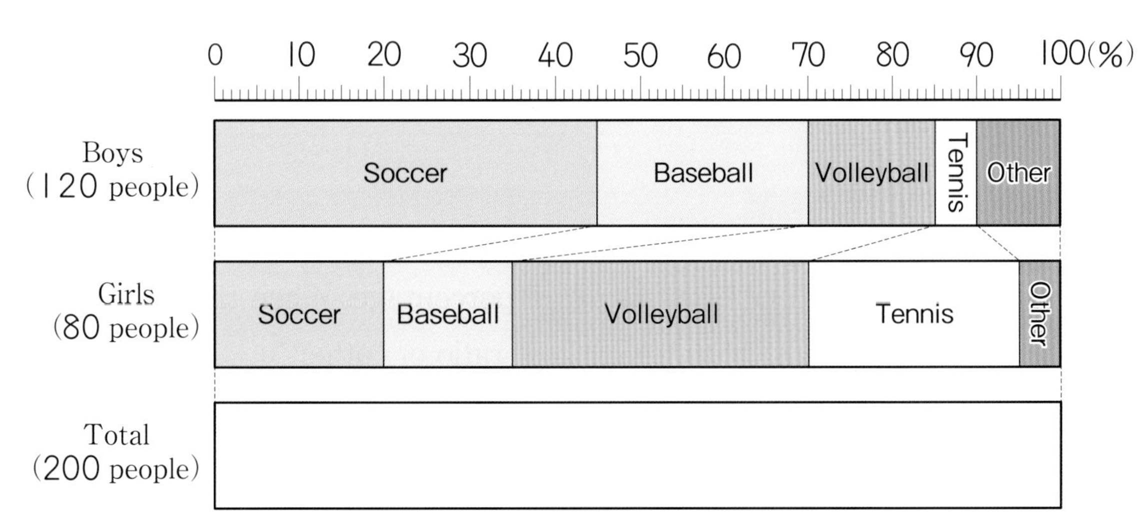


Want to discuss

⑥ Let's discuss what you understood from the above strip graph. Also, let's discuss the differences from a circle graph.

## 3 How to draw circle graphs and strip graphs

Want to represent

**1** **The following tables show the number of elementary school traffic accidents in a city by cause. Let's represent these in a circle graph and strip graph.**

**Number and ratio of traffic accidents by cause (1st Grade)**

| Cause | Number of students | Percentage (%) |
|---|---|---|
| Rush out | 11 | |
| Crossing outside the crosswalk | 4 | |
| Traffic light violation | 3 | |
| Crossing around a car | 3 | |
| Other | 2 | |
| Total | 23 | |

**Number and ratio of traffic accidents by cause (5th Grade)**

| Cause | Number of students | Percentage (%) |
|---|---|---|
| Rush out | 8 | |
| Crossing outside the crosswalk | 9 | |
| Traffic light violation | 4 | |
| Crossing around a car | 2 | |
| Other | 5 | |
| Total | 28 | |

① For the ratio of each cause, let's find the percentage rounding off the thousandths place.

Way to see and think 

Ratio=
compared quantity
÷base quantity.

### How to draw a graph

(1) Find the ratio of each part as a percentage. When the total does not add up to 100, adjust with the largest ratio or "other."

(2) Divide the graph according to the percentages.

(3) Usually, the percentages are drawn from largest to smallest. In a circle graph, the orientation is from top center to the right. In a strip graph, the orientation is from left to right. Additionally, in the case of comparing 2 or more graphs, align the graphs into the same type.

(4) The ratio for "other" is drawn last, even when the percentage is the largest.

② Let's represent with a circle graph.

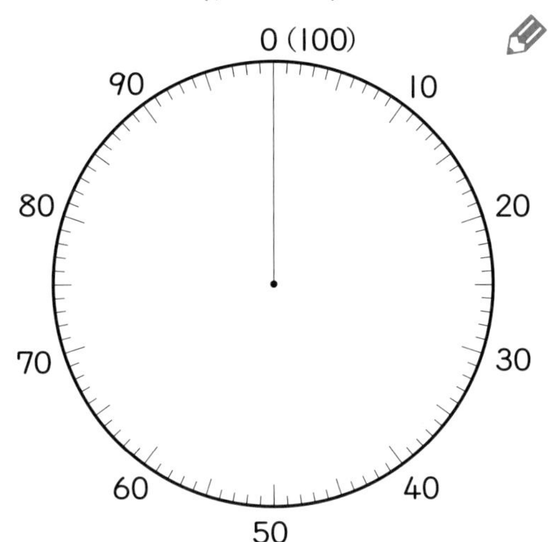


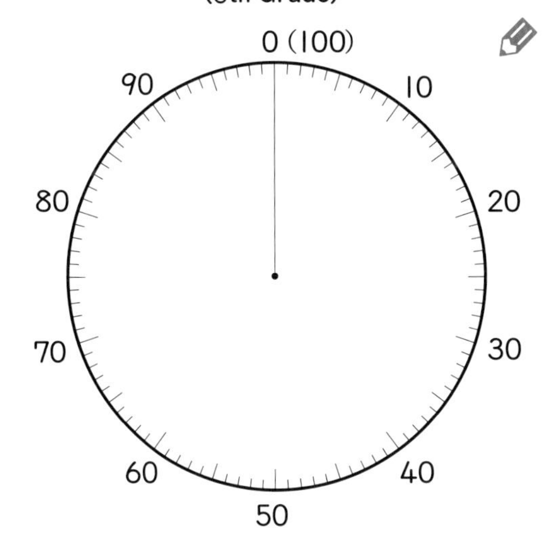


③ Let's represent with a strip graph.

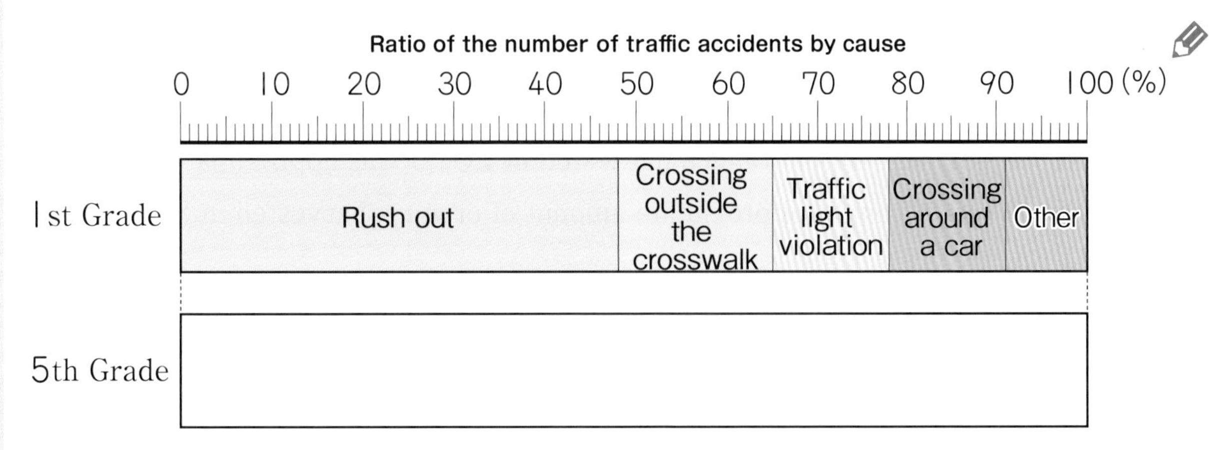


④ Let's look at the above circle graphs and strip graphs, and discuss what you understood.

In the circle graph, it is easy to understand whether the ratio is large or small in that grade.

If you want to compare 1st and 5th grade, the strip graph is easier to understand.

Way to see and think 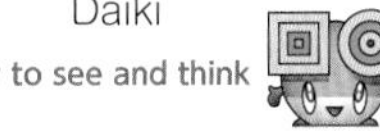

Depending on things you want to know and represent, you should decide the type of graph.

# What you can do now

**Can read and draw a circle graph.**

**1** The circle graphs below show the ratio of amount of kiwis and oranges harvested by prefecture in 2016. Let's answer the following questions.

**Ratio of amount of kiwis harvested by prefecture**

0(100), 10, 20, 30, 40, 50, 60, 70, 80, 90

Ehime, Fukuoka, Wakayama, Kanagawa, Shizuoka, Other

**Ratio of amount of oranges harvested by prefecture**

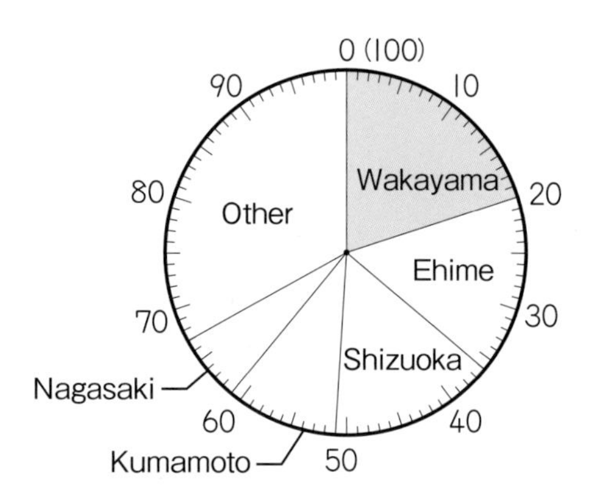


① What percentage of the total is each of the colored parts?

② The national amount of oranges harvested in 2016 was approximately 805100 t. How many tons was the approximate amount of oranges harvested in Shizuoka Prefecture?

③ The approximate amount of kiwis harvested on Wakayama Prefecture in 2016 was 3810 t. Approximately, how many tons was the national amount of kiwis harvested in 2016?

**Can read and draw a strip graph.**

**2** The graph below shows the ratio of number of vehicles that passed in front of the school by type. Let's answer the following questions.

**Vehicles that passed in front of the school**

0, 10, 20, 30, 40, 50, 60, 70, 80, 90, 100(%)

Car, Truck, Bicycle, Bus, Other

① What percentage of the total is the number of cars?

② When the investigation was done, a total of 50 vehicles passed in front of the school. For each type, how many vehicles passed in front of the school?

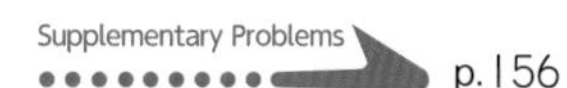

p.156

# Usefulness and efficiency of learning

**1** In Karen's class, students answered which one pet they wanted to own. The circle graph on the right shows the ratio of number of students that want to own a dog based on the total number of students in the class.

① 10 students want to own a dog. How many students is the total number of students in the class?

② The table below summarizes other pets that students want to own. Let's fill in the blank spaces in the table and complete the above circle graph.

**Pets students want to own**

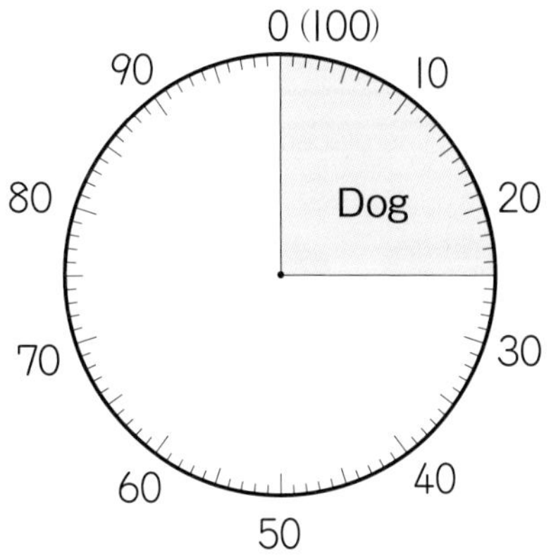

□ Can read and draw a circle graph.

**Pets that students want to own**

| | Dog | Cat | Small bird | Goldfish | Other |
|---|---|---|---|---|---|
| Number of students | | 8 | 2 | 4 | |
| Ratio (%) | | | | | |

**2** The table on the right summarizes the results of exploring the number of Internet uses in a certain town in 2017. Additionally, the strip graph below shows the ratio of number of Internet uses in the same town 10 years ago. By using the results from the table, let's represent the results in 2017 in the following strip graph.

□ Can read and draw a strip graph.

**Ratio of the number of Internet uses in 2017**

| Number of Internet uses | Percentage (%) |
|---|---|
| At least once a day | 76 |
| At least once a week | 16 |
| At least once a month | 5 |
| Less than that | 3 |

**Ratio of the number of Internet uses**

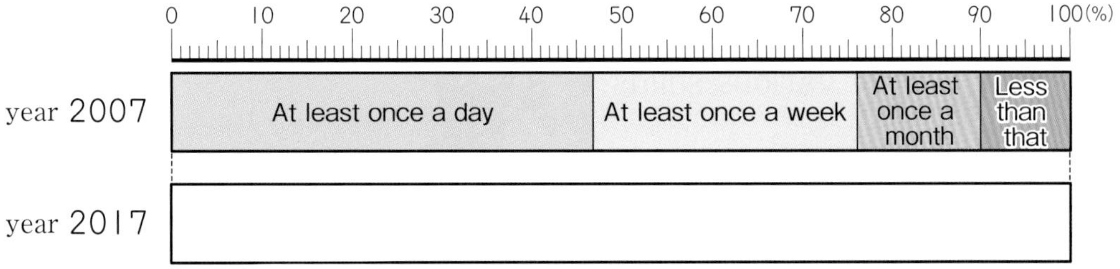

# What kind of shape?

One solid from ⓐ to ⓖ is placed inside the box. Let's guess what kind of shape is inside the box.
ⓐ
ⓑ
ⓒ
ⓓ
ⓔ
ⓕ
ⓖ
1

Don't look inside the box and give three hints that you can find by touching with your hand.
2

It is not a round shape.
It's neither ⓑ nor ⓖ.
3

Some faces are not square or rectangular.
Maybe ⓒ, ⓓ, or ⓕ?
4

It's ⓕ!
6


**Problem** What are the properties of various solids?

# 19 Solids

# Let's explore the properties of various solids.

## 1 Prisms and cylinders

Want to classify

Activity

**1** Let's classify the solids ⓐ to ⓖ into two groups.

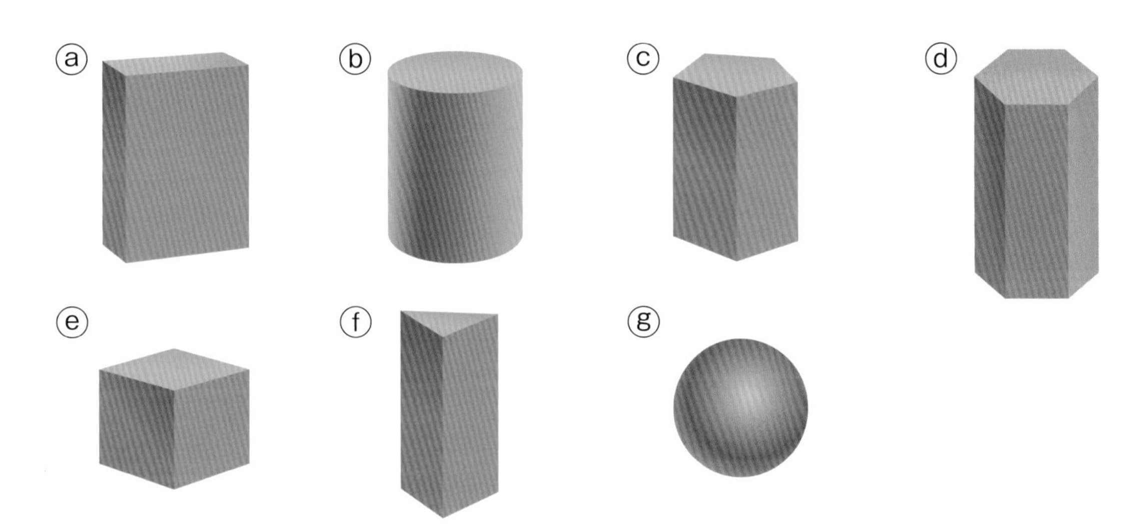


The face that bends and is not plane is called a **curved face**. The shape that is covered by planes or curved faces is called a **solid**.

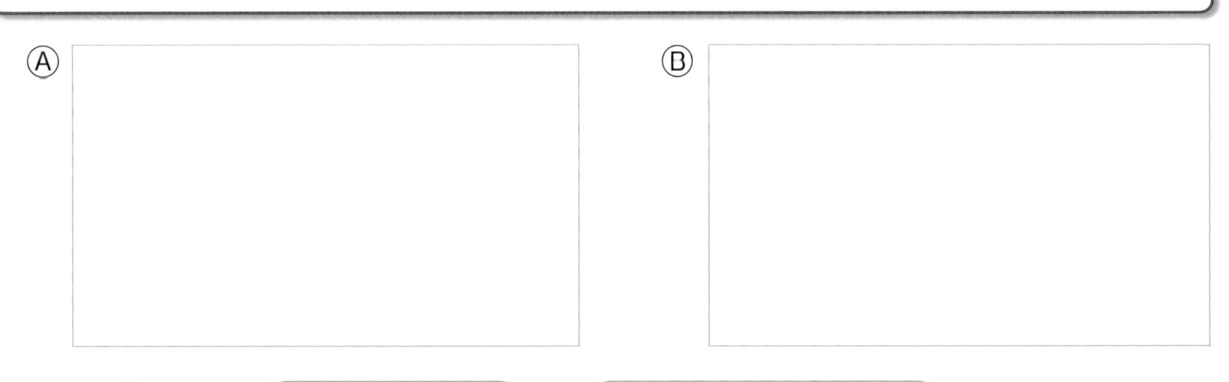

Daiki: Where should I focus to classify?

Yui: Some have curved faces, others do not.

Want to explain

① Yui classified the solids as shown below. What are the features of each of the solids in Ⓐ and the solids in Ⓑ?

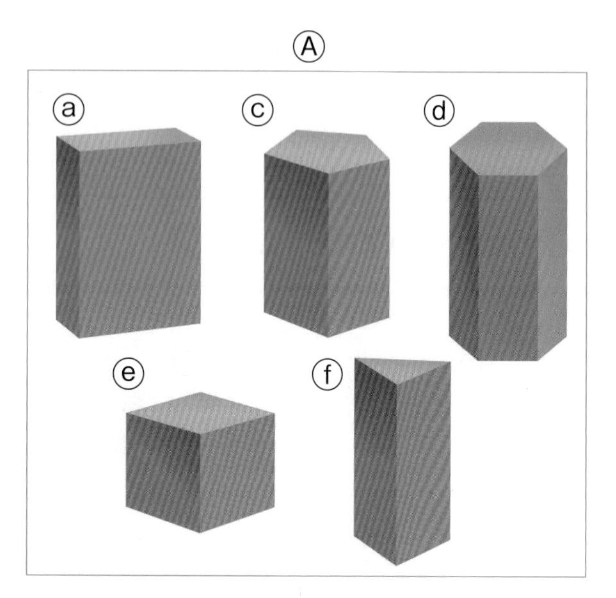


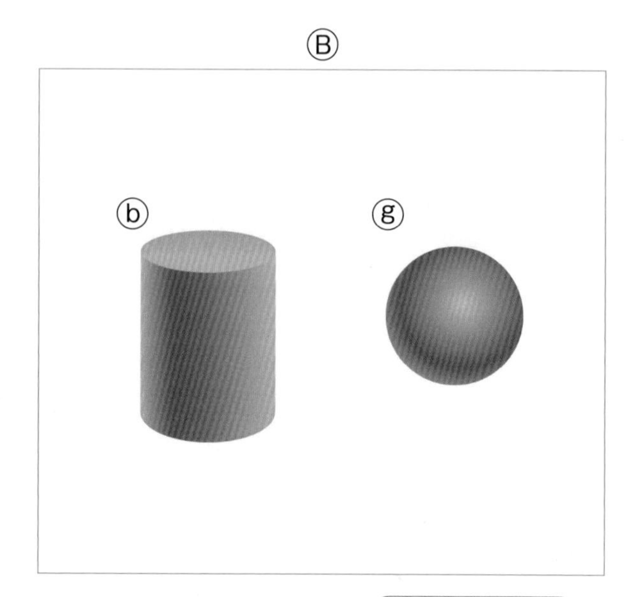

Such solids as ⓐ, ⓒ, ⓓ, ⓔ, and ⓕ are called **prisms**.
A solid such as ⓑ is called a **cylinder**.

ⓖ is a sphere.

Hiroto

Want to explore Properties of prisms

## 2 Let's explore the following prisms Ⓐ to Ⓓ.

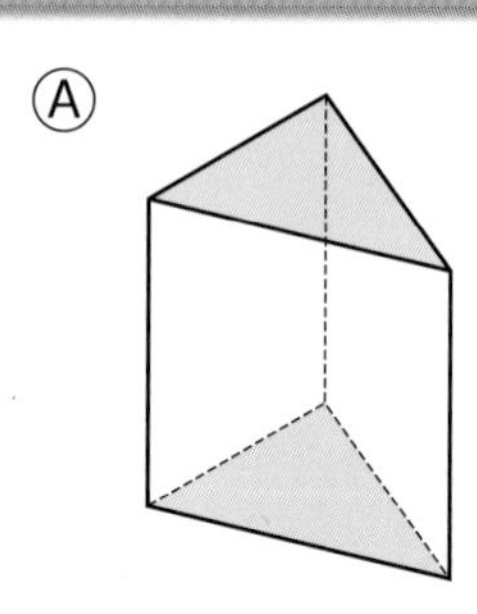


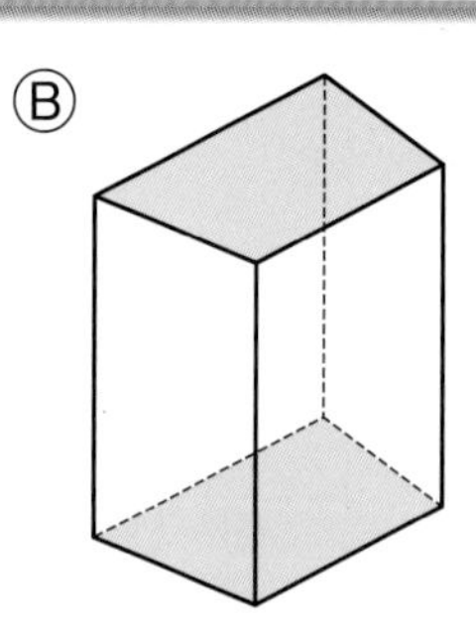


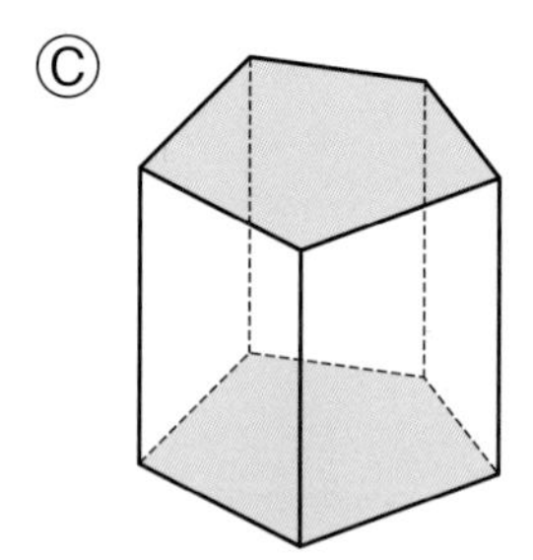


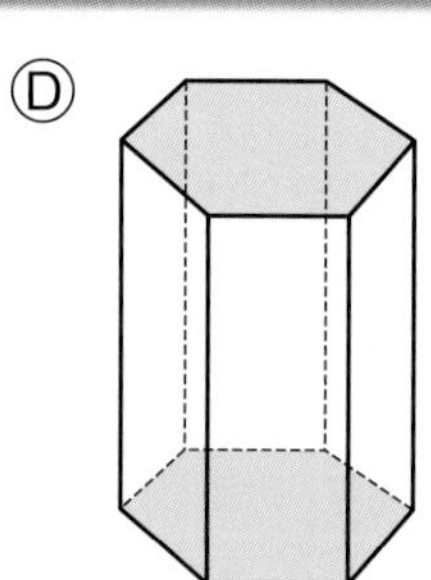


① In each solid, the pair of colored faces are parallel. What is the shape of the colored faces? Also, is each pair of shapes congruent?

② What is the shape of the faces that are not colored? Also, how many faces are there?

③ Which faces are perpendicular?

The two parallel congruent faces of a prism are called **bases**, and the rectangular or square faces around the bases are called **lateral faces**. When the bases are triangles, quadrilaterals, pentagons, ...., these prisms are called **triangular prism**, **quadrangular prism**, **pentagonal prism**, ...., respectively.

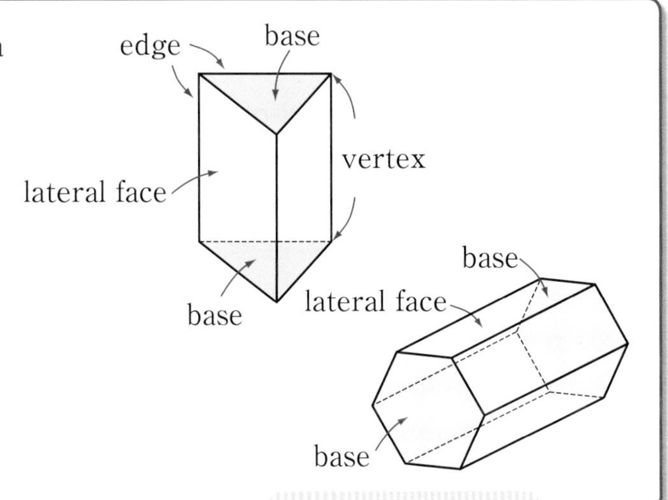


④ Let's say the names of the solids Ⓐ to Ⓓ.

You can see the cubes and cuboids as quadrangular prisms.

Want to summarize

⑤ Let's summarize the faces, vertices, and edges of prisms in the table.

| | Triangular prism | Quadrangular prism | Pentagonal prism | Hexagonal prism |
|---|---|---|---|---|
| Shape of bases | Triangle | | | |
| Shape of lateral faces | Rectangle or square | | | |
| Number of faces | | | | |
| Number of vertices | | | | |
| Number of edges | | | | |

Way to see and think

If you try to summarize, it is easier to understand what is common.

⑥ Daiki found the number of faces, vertices, and edges of a triangular prism as shown on the right. Let's explain Daiki's idea.

Daiki's idea

Number of faces $2 + 3 = 5$

Number of vertices $3 \times 2 = 6$

Number of edges $3 \times 2 + 3 = 9$

⑦ Let's look at the table in ⑤ and discuss what kind of rules there are.

Want to explore Properties of cylinders

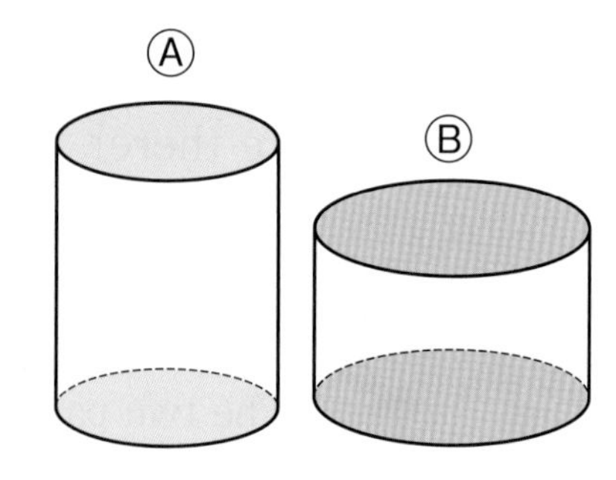


## 3 Let's explore cylinders Ⓐ and Ⓑ shown on the right.

Want to find

① What kind of faces are they covered with?

② What kind of shape are the parallel faces? Also, are they congruent?

The two parallel congruent circles of a cylinder are called **bases** and the curved face around the bases is called the **lateral face**.

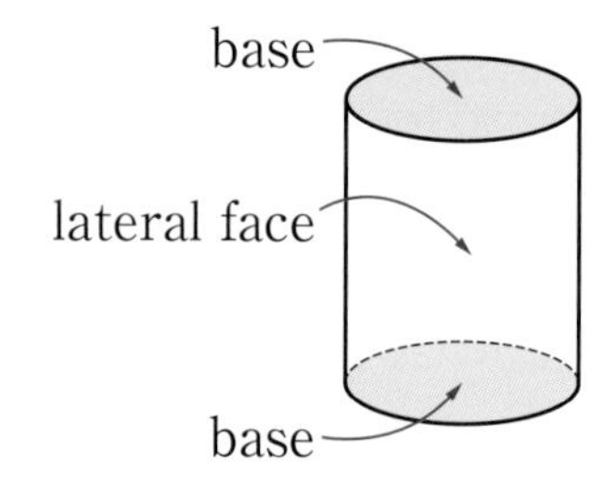


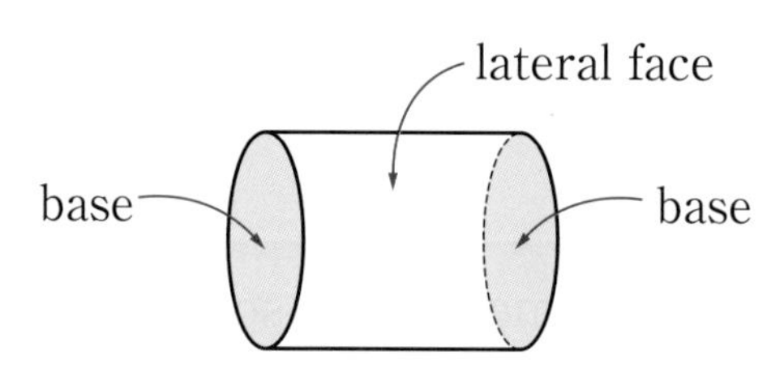


The length of the line that is perpendicular to the two bases of a prism or a cylinder is called the **height** of the prism or cylinder, respectively.

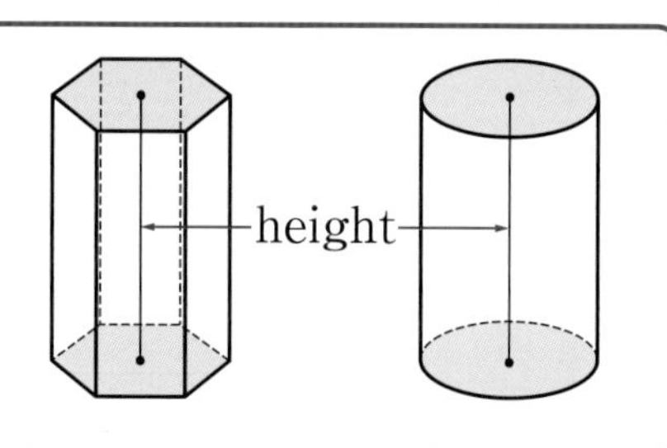


Want to confirm

Let's answer the following questions about the solid shown on the right.

① What kind of solid is this?

② Which faces are parallel and perpendicular to face ABCD?

③ Which sides shall we use for measuring the height of this solid? Let's write them all.

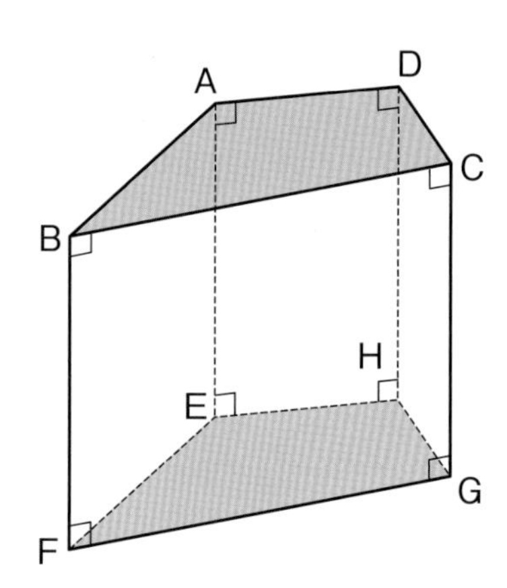

## Sketch and net

Want to represent Sketch

**1 Let's draw sketches of the following triangular prism and cylinder.**

① Triangular prism

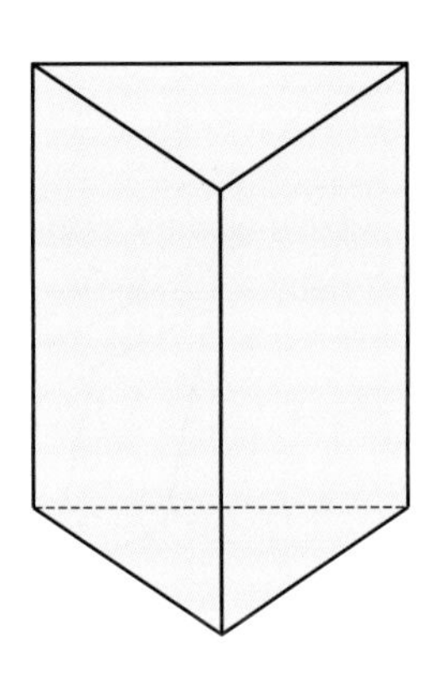

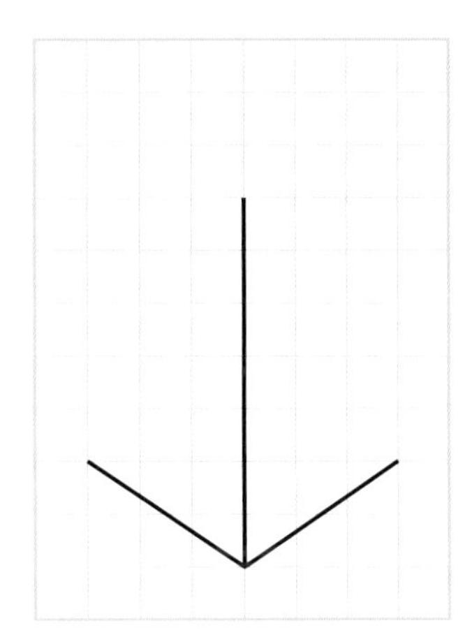

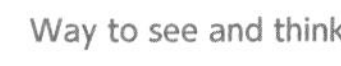

Remember how to draw the sketches of cuboids and cubes, and try to draw in the same way.

Daiki

In the sketch, parallel edges are drawn parallel.

Is it okay if I draw dotted lines for invisible edges?

Yui

② Cylinder

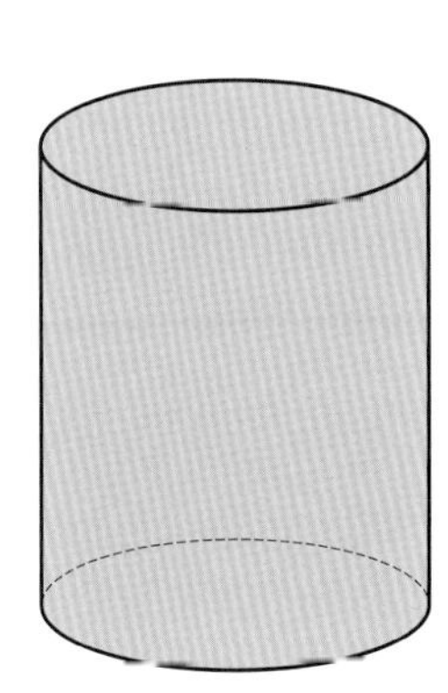

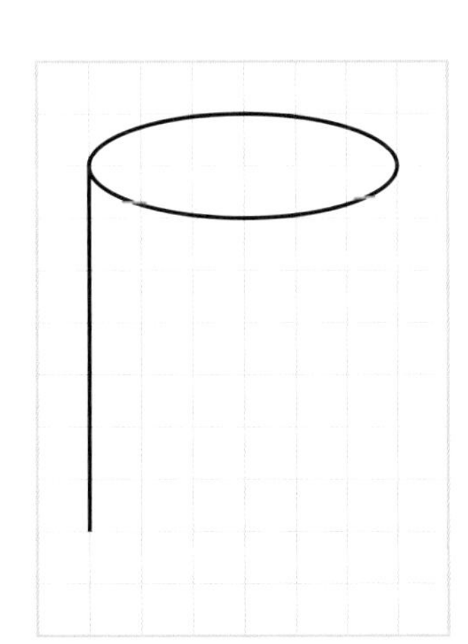

Want to represent  Net of a prism

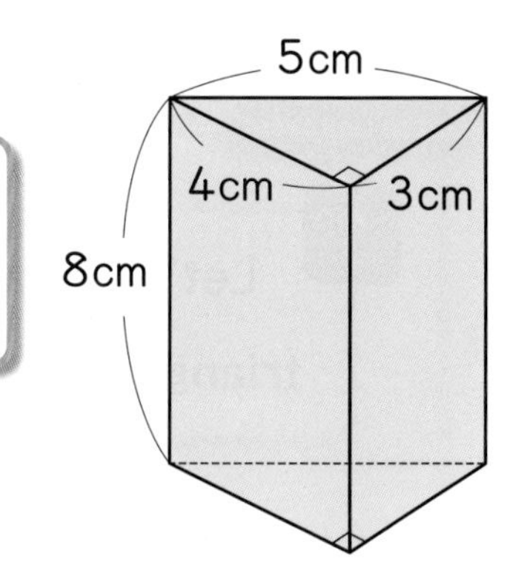


## 2 Let's draw a net in the cardboard to assemble the triangular prism shown on the right.

1cm
1cm

J
A K H G
B C E F
D

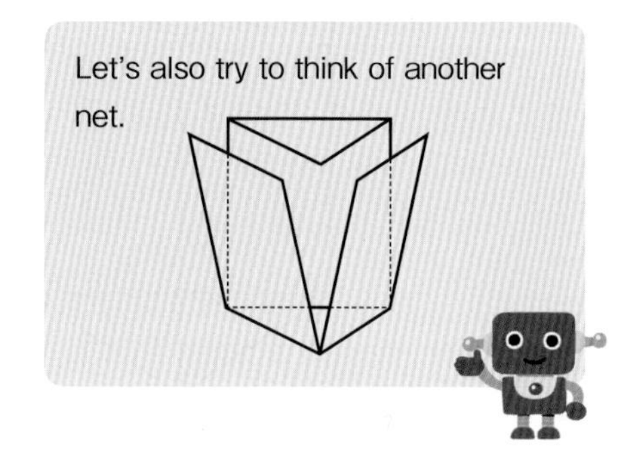

Let's also try to think of another net.


① Which parts of the net are the bases and the lateral faces of the prism?

② Which part of the net is the height of the triangular prism?

③ How many cm is the length of side AB, side BC, and side DE?

④ When you assemble the prism, which points gather at point A?

Want to confirm

Let's make the triangular prism shown on the right by drawing a net.

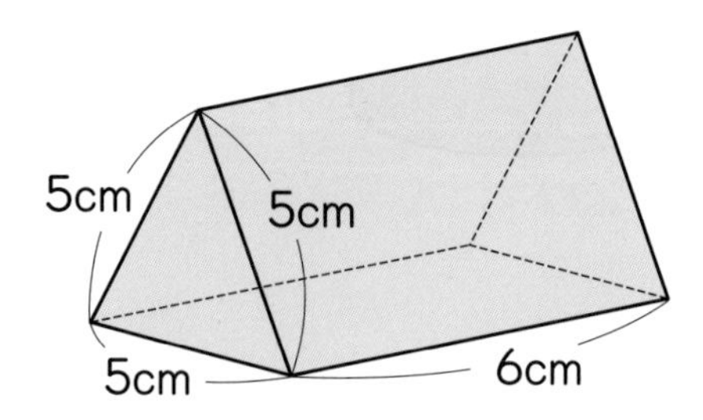

Want to represent Net of a cylinder

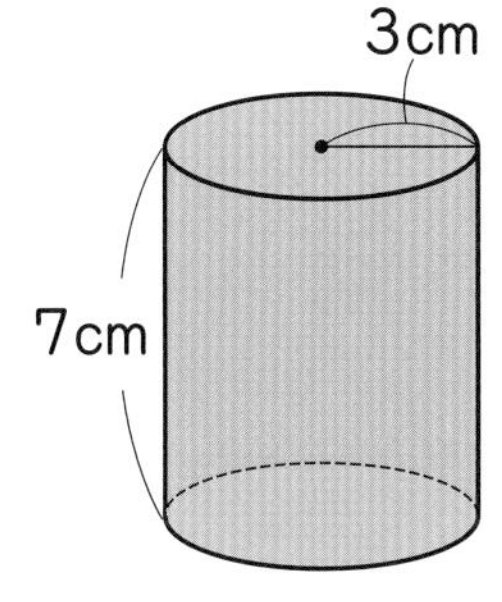


**3** **Let's draw a net in the cardboard to assemble the cylinder shown on the right.**

Since the two bases are circles, I can draw them but...

What should I do to draw the lateral face?

Purpose What kind of shape is the net of a cylinder?

① First, roll a sheet of paper around the lateral face as shown on the right. Then, open the paper to draw the net. What is the shape of the lateral face?

② Which part of the net, shown on the right, is equal to the height of the cylinder?

③ Which part of the base is equal to the length of straight line AD?

④ Let's find the length of straight line AD.

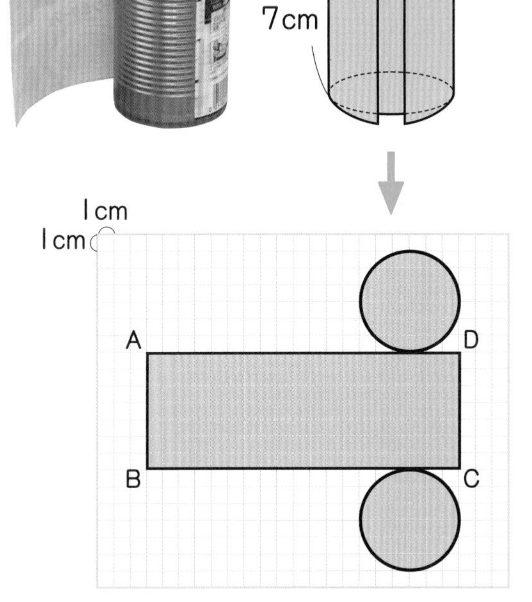


Summary

The net of the lateral face of a cylinder is a rectangle. The length is equal to the height of the cylinder and the width is equal to the circumference of a base.

Want to confirm

Let's make the cylinder shown on the right by drawing a net.

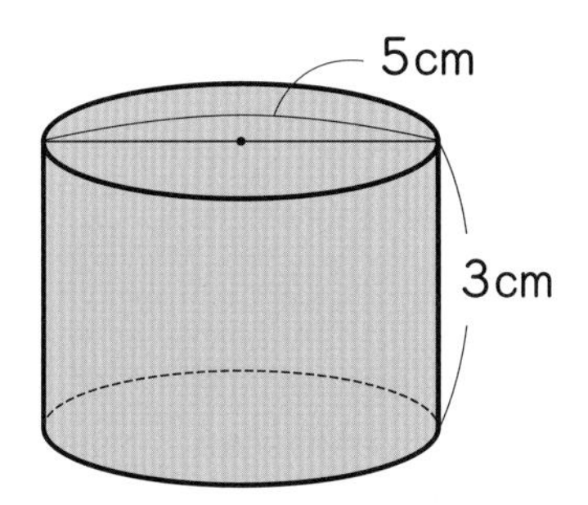

# What you can do now

□ **Understanding the properties of prisms.**

**1** A solid is shown on the right. Let's answer the following questions.

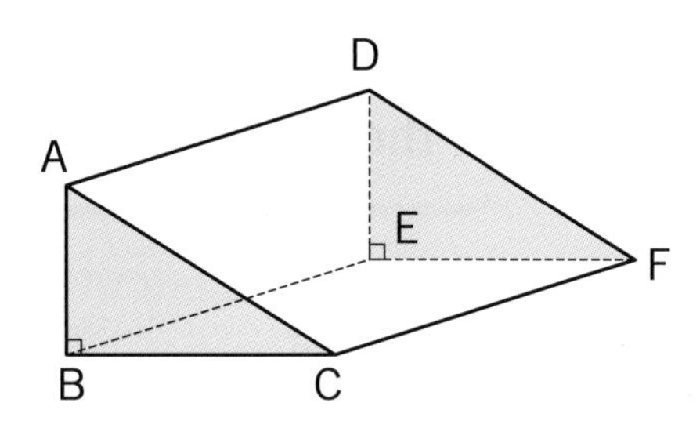


① What kind of solid is this?

② How many faces and edges does it have?

③ Which faces are parallel and perpendicular to face ABC?

④ What sides shall we use for measuring the height of this solid?

□ **Understanding the properties of cylinders.**

**2** Let's explore the solid shown on the right.

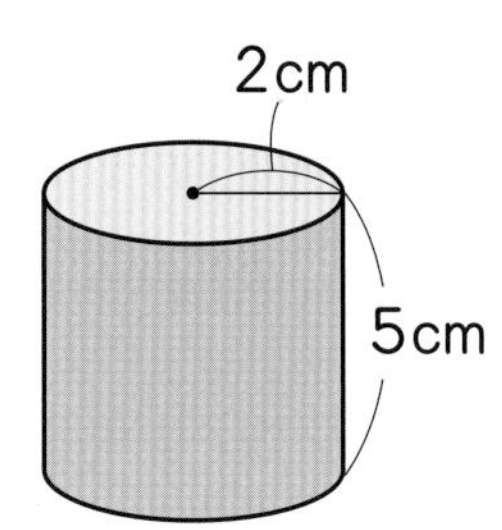


① What kind of solid is this?

② How many cm is the height?

□ **Understanding about sketches and nets.**

**3** Let's answer the following questions.

① What kind of solid is formed if the following nets are assembled?

Ⓐ

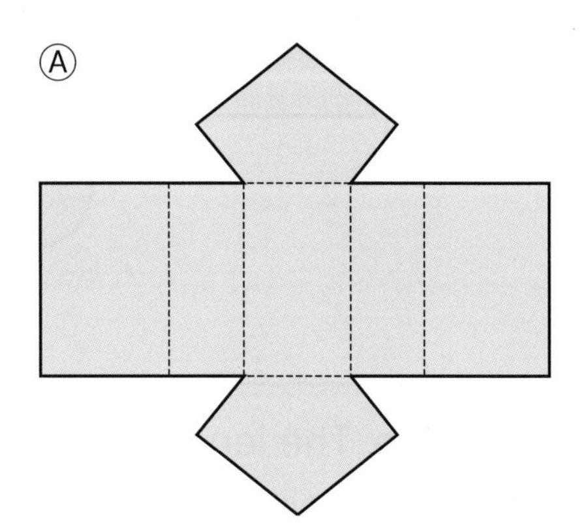

Ⓑ

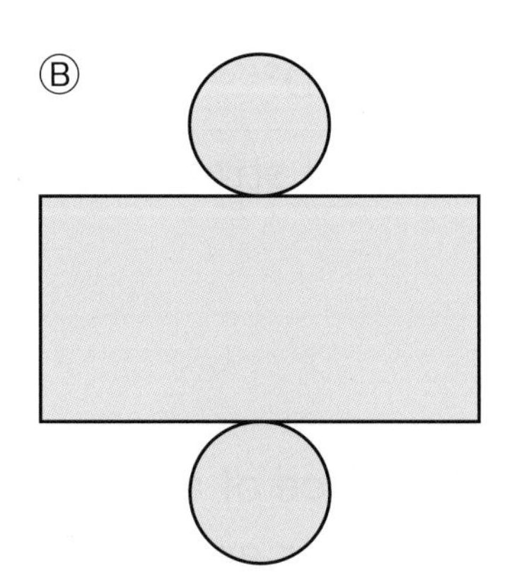

Ⓒ

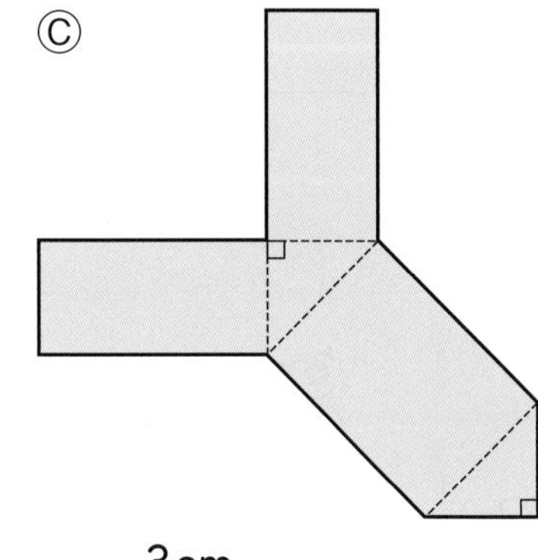

② Let's find the width of the lateral face when the net is drawn. Use 3.14 as the ratio of circumference. Let's find the nearest tenths place round number by rounding off the hundredths place.

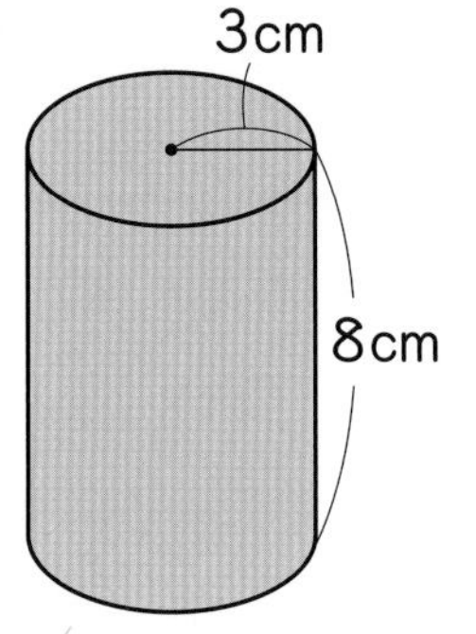


Supplementary Problems p.157

**1** Let's summarize what we know about prisms in the following table.

☐ Understanding the properties of prisms.

| | Heptagonal prism | Octagonal prism | Nonagonal prism | Decagonal prism |
|---|---|---|---|---|
| Number of faces | | | | |
| Number of vertices | | | | |
| Number of edges | | | | |

**2** The diagram on the right is the net of a triangular prism. Let's answer the following questions.

☐ Understanding about sketches and nets.

① How many cm is the height of the assembled triangular prism?

② How many cm is the length of the following sides?

Ⓐ side AB

Ⓑ side FG

③ When you assemble the prism, which points gather at the following points?

Ⓐ point A

Ⓑ point D

④ Let's draw a sketch of the assembled triangular prism in your notebook.

**3** Using a rectangular cardboard as shown on the right, we will make a cylinder by placing sides AB and DC edge to edge.

To make the bases, how many centimeters should the diameters of the circles be?

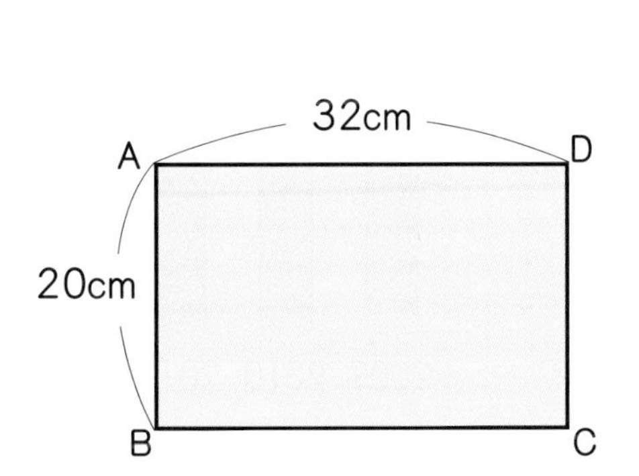


☐ Understanding the properties of cylinders.

☐ Understanding about sketches and nets.

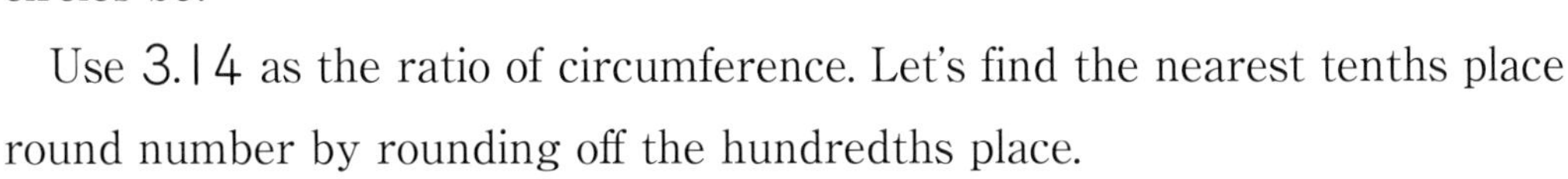
Use 3.14 as the ratio of circumference. Let's find the nearest tenths place round number by rounding off the hundredths place.

# 20 Utilization of data

## Let's interpret trends in the data from graphs.

Want to know

**1** **The following graph shows the number of rented books and the number of visitors (number of people who came to the library) of a library from 2009 to 2017. The bar graph shows the number of rented books and the line graph shows the number of visitors. Let's answer the questions on the next page.**

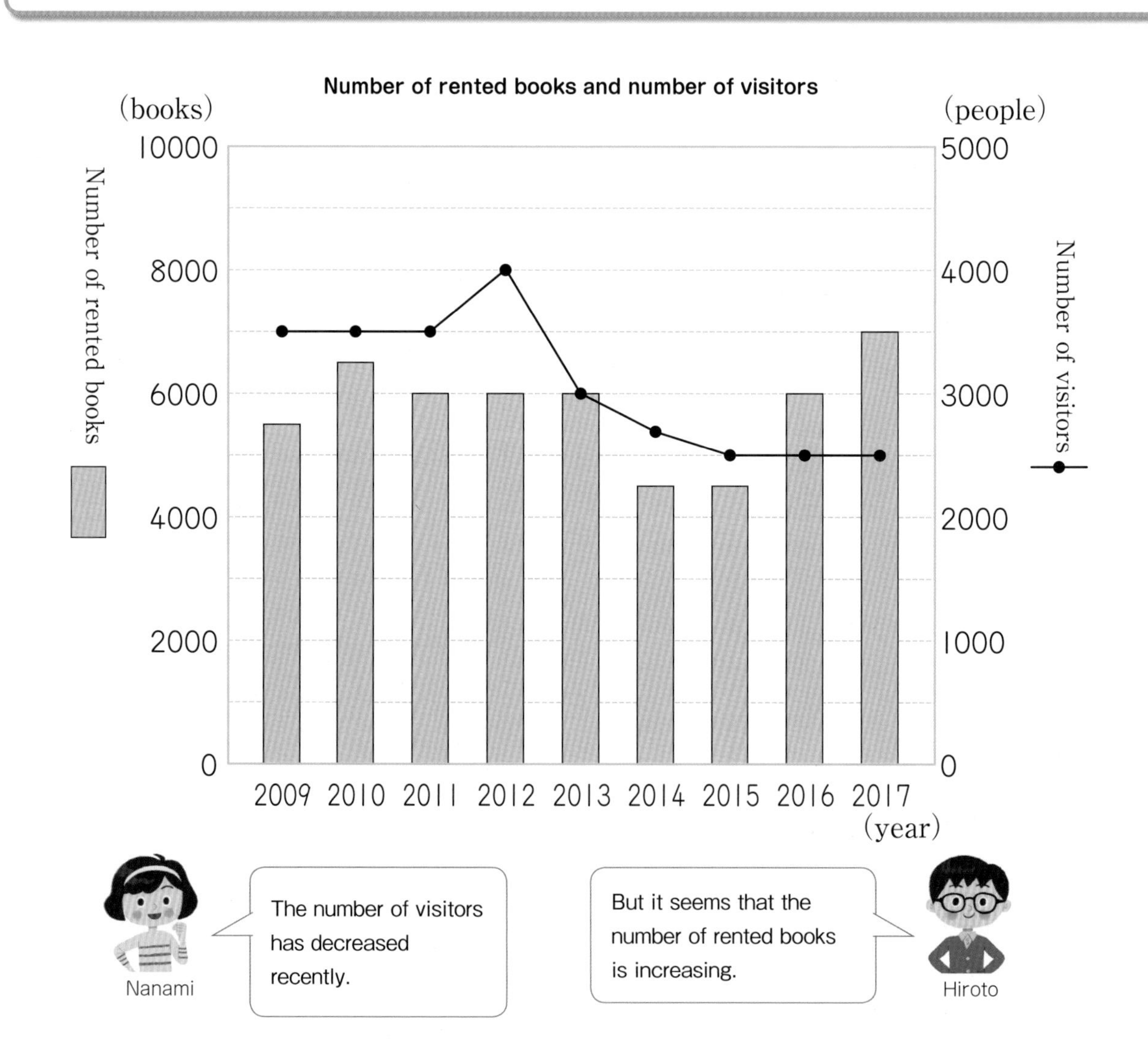

Want to find

① Daiki and Yui looked at the graph and thought as follows. The period considered by each child is from what year to what year? Let's choose from the following Ⓐ to Ⓓ.

Ⓐ From 2009 to 2011.　Ⓑ From 2011 to 2013.

Ⓒ From 2013 to 2015.　Ⓓ From 2015 to 2017.

Daiki's idea

There is a period when the number of visitors is increasing or decreasing, but the number of rented books is unchanged.

Yui's idea

Although the number of visitors has not changed, there is a period when the number of books is increasing.

Want to compare

② Recently, the ratio of rented books using the Internet has been increasing. The following strip graphs show the total number of rented books in the library and the ratio of the number of rented books using the Internet from 2014 to 2017.

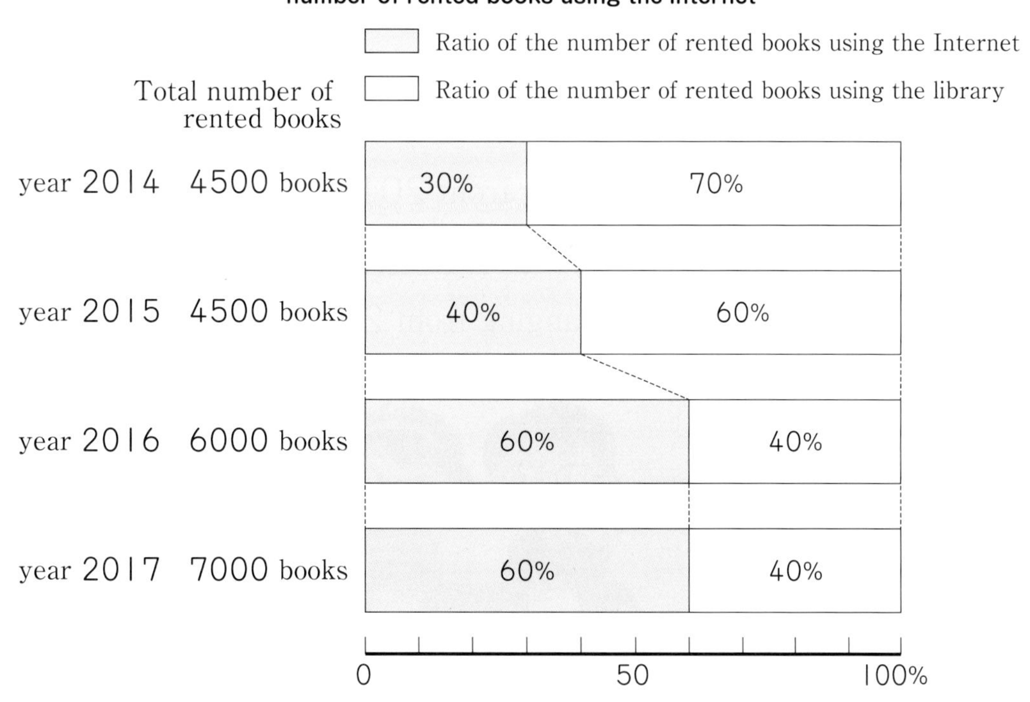

Daiki and Yui compared years 2014 and 2015 from the graphs in the previous page. Let's compare their ideas.

Daiki's idea

30 % and 40 % are represented in decimal numbers as 0.3 and 0.4.
Since $4500 \times 0.3 = 1350$,
then the number of books in year 2014 is 1350 books.
Since $4500 \times 0.4 = 1800$,
then the number of books in year 2015 is 1800 books.
Therefore, the number of rented books using the Internet is larger in 2015.

| Base quantity | Compared quantity |
|---|---|
| 4500 books | □ books |
| 1 | 0.3 |
| | Ratio |

| Base quantity | Compared quantity |
|---|---|
| 4500 books | □ books |
| 1 | 0.4 |
| | Ratio |

Yui's idea

The ratio of rented books using the Internet was 30 % in 2014 and 40 % in 2015. Since the number of rented books is the same, the number of rented books using the Internet is larger in 2015 because the ratio is larger.

③ If you compare years 2016 and 2017, is the number of rented books using the Internet increasing? Let's choose one from the following Ⓐ, Ⓑ, and Ⓒ.

Also, write down the reasons for your choice by using words, numbers, and math expressions based on either of the children's ideas.

Ⓐ The number is increasing from 2016 to 2017.

Ⓑ The number is decreasing from 2016 to 2017.

Ⓒ The number is not changing from 2016 to 2017.

Just find the number of books and you can understand which is increasing.

It looks like you can compare without having to find the number of books.

**2** The table below summarizes the amounts of consumption, import rates, and imports of mineral water from 2000 to 2016. "Amount of consumption" represents the total amount produced in Japan and imported from abroad, and the "Import rate" indicates the ratio of the amount of consumption that is imported. Let's answer the following questions about this table.

Amount of consumption, import rate, and import of mineral water

| Year | 2000 | 2005 | 2010 | 2011 | 2012 | 2013 | 2014 | 2015 | 2016 |
|---|---|---|---|---|---|---|---|---|---|
| Amount of consumption (10000 kL) | 109.0 | 183.4 | 251.8 | 317.2 | 314.1 | 325.5 | 326.1 | 338.7 | 352.3 |
| Import rate (%) | 17.9 | 22.2 | 16.6 | 18.6 | 11.2 | 12.0 | 10.5 | 10.3 | 9.8 |
| Amount of import (10000 kL) | 19.5 | 40.7 | 41.8 | 59.0 | 35.2 | 39.0 | 34.3 | 34.9 | 34.6 |

Amount of consumption and import rate of mineral water

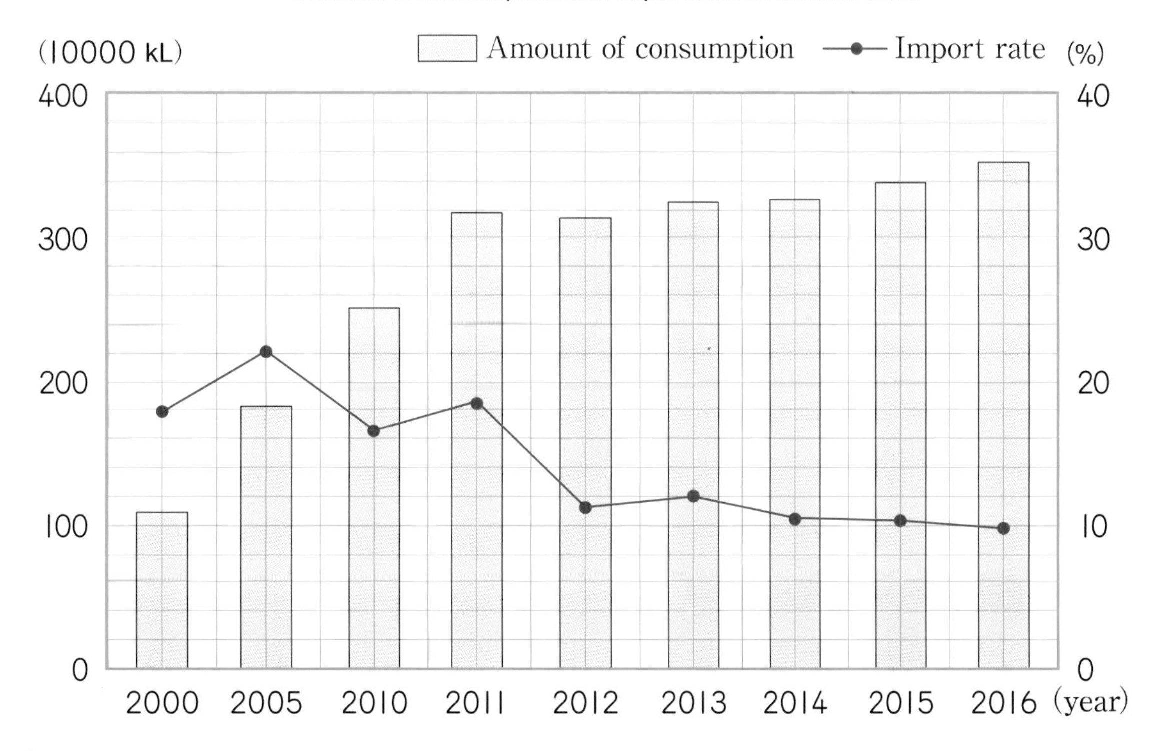


① Let's look at the above table and graph, and discuss what you noticed.

② Nanami thought the following from the graph in the previous page. Is this idea correct?

Nanami's idea

From the graph, it can be said that the increase in the import rate between the years 2000 and 2005 is almost the same as the increase in the import rate between the years 2010 and 2011.

③ Hiroto drew the graph from the previous page as follows. Let's explain how he changed it?

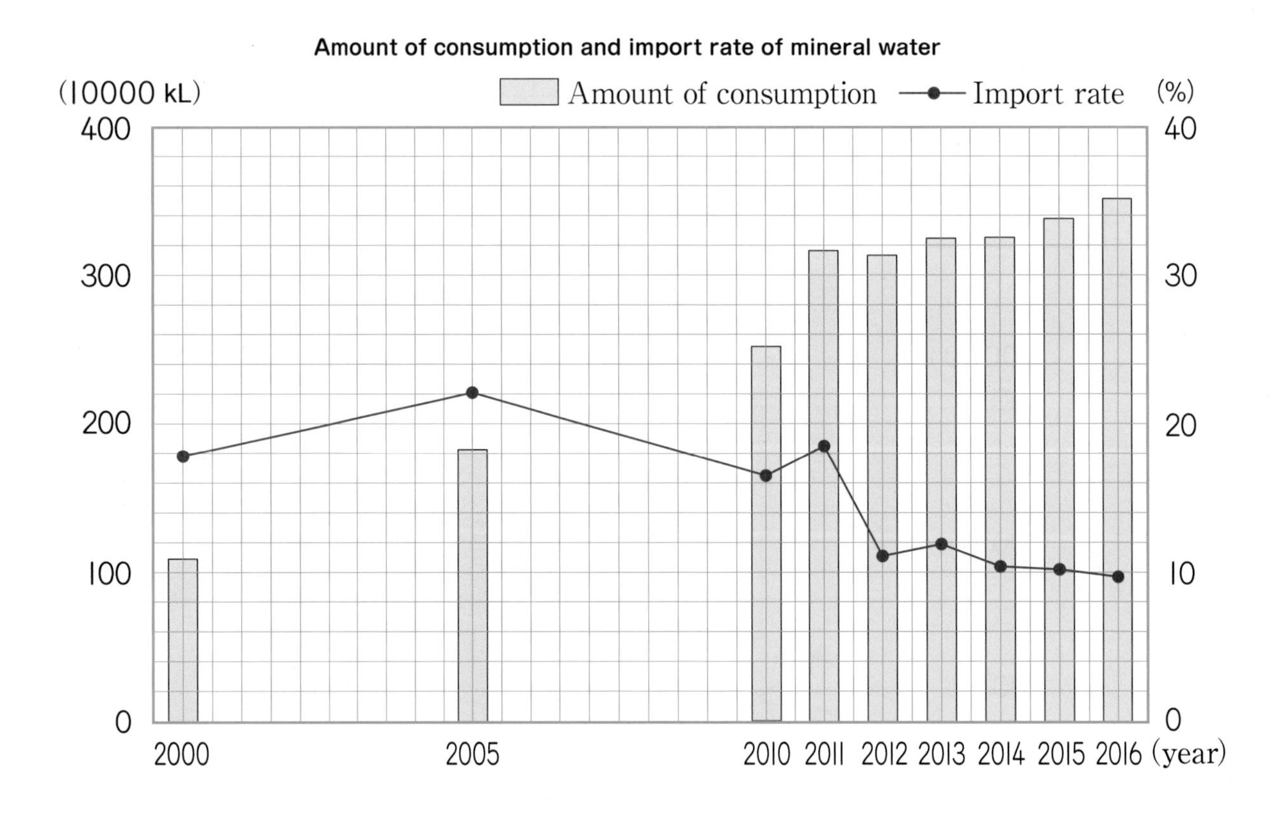


Develop in Junior High School

From 2, the amount of consumption increases by about how many ten thousand kL per year between 2014 and 2016? Also, assuming that the trend continues, let's estimate about how many ten thousand kL will be consumed in 2020 and 2025.

## PPDAC cycle.

When you find a problem in your surroundings that you want to solve, one method to solve it is called the PPDAC cycle. Let's consider the problem in page 135 as an example to see how the cycle might look like.

**P** (Problem)…finding a problem.

On the problem from page 135, we found and explored how much mineral water is being consumed and how much is imported from abroad.

**P** (Plan)…making a plan.

The plan was to explore how the amount of consumption and import rate of mineral water are changing, and to understand the trends.

**D** (Data)…collecting data.

Using the Internet, we collected data on the amount of consumption, import rate, and import of mineral water.

Amount of consumption, import rate, and import of mineral water

| Year | 2000 | 2005 | 2010 | 2011 | 2012 | 2013 | 2014 | 2015 | 2016 |
|---|---|---|---|---|---|---|---|---|---|
| Amount of consumption (10000 kL) | 109.0 | 183.4 | 251.8 | 317.2 | 314.1 | 325.5 | 326.1 | 338.7 | 352.3 |
| Import rate (%) | 17.9 | 22.2 | 16.6 | 18.6 | 11.2 | 12.0 | 10.5 | 10.3 | 9.8 |
| Amount of import (10000 kL) | 19.5 | 40.7 | 41.8 | 59.0 | 35.2 | 39.0 | 34.3 | 34.9 | 34.6 |

Amount of consumption and import rate of mineral water

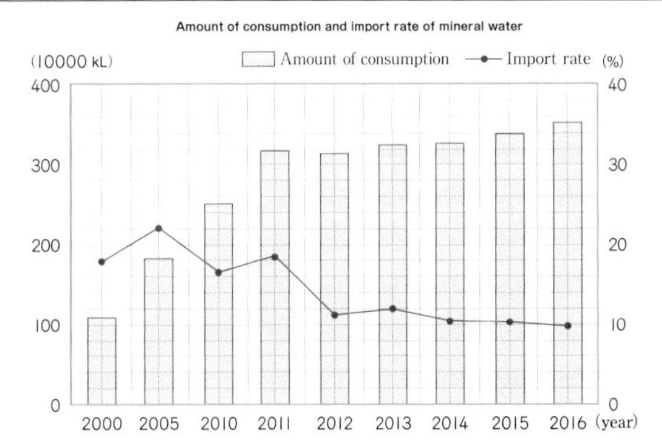


**A** (Analysis)…analyzing data.

We represented the data in tables and graphs according to the purpose, analyzed the data and found out what we could understand.

**C** (Conclusion)…reaching a conclusion.

From what we explored until now, we concluded what we could understand about the amount of consumption and import rate of mineral water.

After you have reached a conclusion from the original problem and found a new problem, the new problem will be solved once more with the PPDAC cycle.

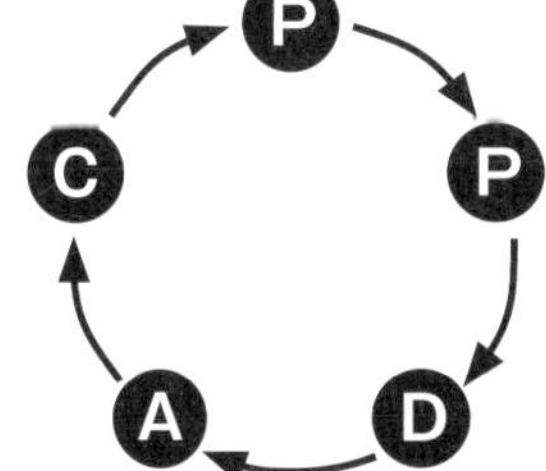


**P** (Problem)…finding a new problem.

From the trends so far, we have found the problem of how consumption will increase in the future.

## 21 5th Grade summary
# Let's review learned mathematics.

Decimal numbers and whole numbers

**1** Let's find the numbers that are 100 times and $\frac{1}{100}$ of the following numbers.

① 5.18 ② 0.407 ③ 13.4 ④ 3600

Multiplication of decimal numbers, division of decimal numbers, and addition and subtraction of fractions

**2** Let's solve the following calculations.

① $8 \times 1.6$ ② $5 \times 2.2$ ③ $32 \times 6.4$

④ $3.8 \times 2.5$ ⑤ $5.72 \times 8.1$ ⑥ $0.4 \times 0.28$

⑦ $9 \div 0.5$ ⑧ $48 \div 1.6$ ⑨ $54 \div 1.8$

⑩ $1.2 \div 0.3$ ⑪ $8.05 \div 3.5$ ⑫ $0.03 \div 0.15$

⑬ $\frac{3}{4}+\frac{1}{8}$ ⑭ $2\frac{1}{8}+1\frac{5}{12}$ ⑮ $1\frac{5}{6}+3\frac{9}{14}$

⑯ $\frac{5}{6}-\frac{2}{3}$ ⑰ $3\frac{8}{15}-1\frac{4}{9}$ ⑱ $3\frac{3}{16}-1\frac{7}{8}$

Multiples and divisors

**3** Let's summarize about the properties of whole numbers.

① In the whole numbers between 50 and 100, how many common multiples of 4 and 6 are there?

② Let's find the multiples and divisors of the following numbers. However, for the multiples, let's find three in ascending order.

Ⓐ 48 Ⓑ 63

③ Let's find the least common multiple and the greatest common divisor for each of the following pair of numbers.

Ⓐ (18, 36) Ⓑ (15, 18)

Fractions, decimal numbers, and whole numbers

**4** Let's arrange the following fractions and decimal numbers in descending order.

① $\frac{4}{5}$ $\frac{17}{8}$ 0.7 1.6 $1\frac{3}{4}$

② 1 0.71 $\frac{2}{3}$ $\frac{3}{5}$ $\frac{7}{10}$

Proportion

**5** The relationship between the weight and length of a wire that weighs 3 g per meter was explored. Let's answer the following questions.

**Weight and length of a wire**

| Length (m) | 1 | 2 | 3 | 4 | 5 | 6 | |
|---|---|---|---|---|---|---|---|
| Weight (g) | 3 | | 9 | | | | |

① Let's write the numbers that apply to the weight in the above table.

② Let's consider length as □ m and weight as ○ g, and write a math sentence to find the weight.

③ When the weight is 60 g, how many meters is the length of the wire?

Mean

**6** The following table shows the number of students and number of times that 5th graders can do pull-ups at Genya's school. Based on this, what is the mean number of times that a 5th grade student can do pull-ups?

**Number of students and number of times that 5th grade students can do pull-ups**

| Number of times | 0 | 1 | 2 | 3 | 4 | 5 | 6 | 7 | 8 | 9 | 10 |
|---|---|---|---|---|---|---|---|---|---|---|---|
| Number of students | 3 | 0 | 2 | 4 | 5 | 16 | 9 | 10 | 4 | 6 | 1 |

Measure per unit quantity and volume

**7** There is a water tap that pours 20 L of water in 4 minutes. How many minutes does it take to fill a water tank with a length of 1 m, width of 50 cm, and depth of 50 cm?

Measure per unit quantity

**8** 966 children are playing in a playground with an area of 1680 $m^2$. 105 children are playing in a court yard with an area of 200 $m^2$. Which of the two places is more crowded?

Measure per unit quantity

**9** Let's summarize about speed.

① Let's write the relationship between speed, distance, and time in a math sentence.

② A person is walking toward a place that is 8 km away. If this person walks at 4 km per hour, how many kilometers are left to the destination after walking for 1.5 hours?

Ratio

**10** Let's find the numbers that apply in the following □.

① 36 kg is □ % of 48 kg.

② 80% of 2.5 m is □ m.

③ 35% of □ yen is 1400 yen.

Ratio

**11** The East Supermarket sells lunch with a discount of 120 yen on Sundays. The West Supermarket sells lunch with a discount of 20 % on Sundays. Let's answer the following questions.

① When you buy a lunch for 500 yen, which supermarket is better?

② When you buy a lunch for 800 yen, which supermarket is better?

③ How many yen is the cost of the lunch that has the same price at the East Supermarket and West Supermarket?

Congruent figures

## 12 From the following figures, which are congruent?

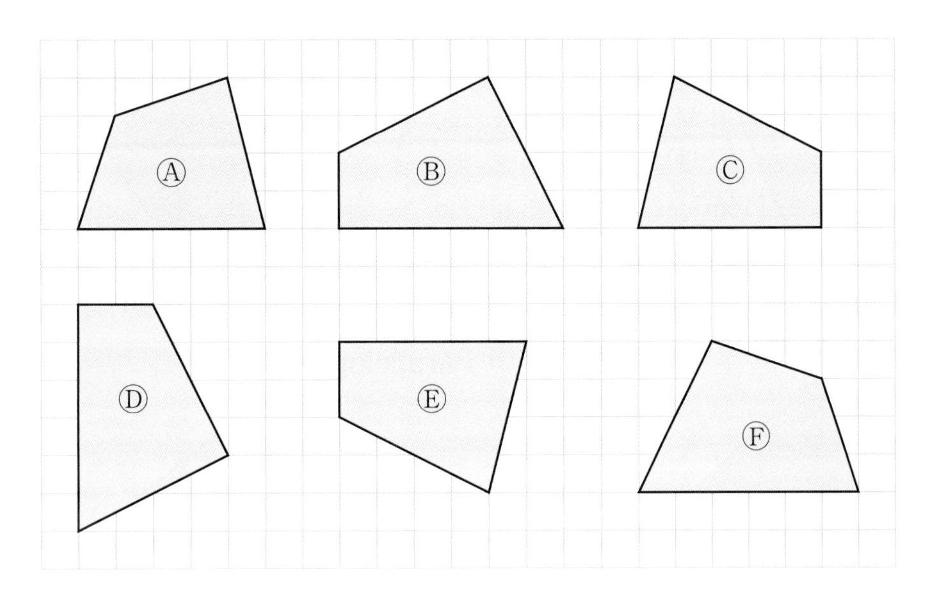


Angles of figures

## 13 Let's find the size of the angles Ⓐ, Ⓑ, Ⓒ, and Ⓓ by calculations.

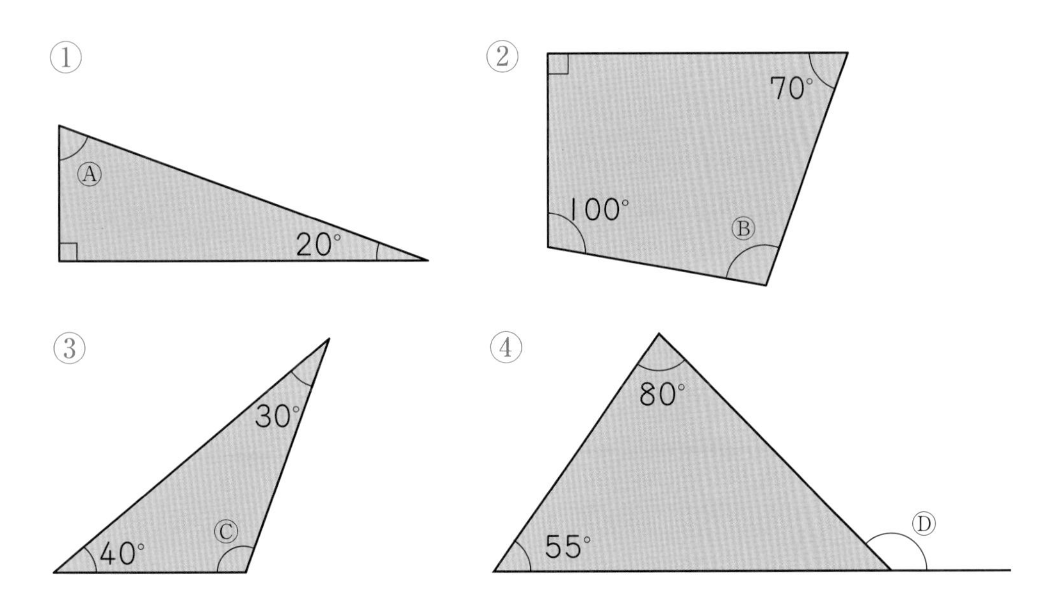

Area of figures

## 14 Let's find the area of the following figures.

①

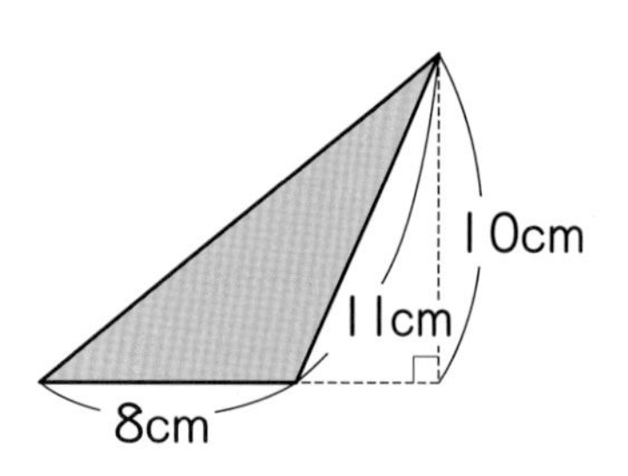


② Parallelogram

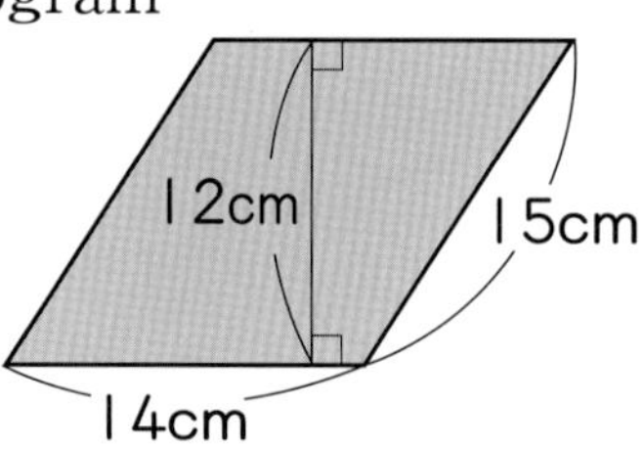


③

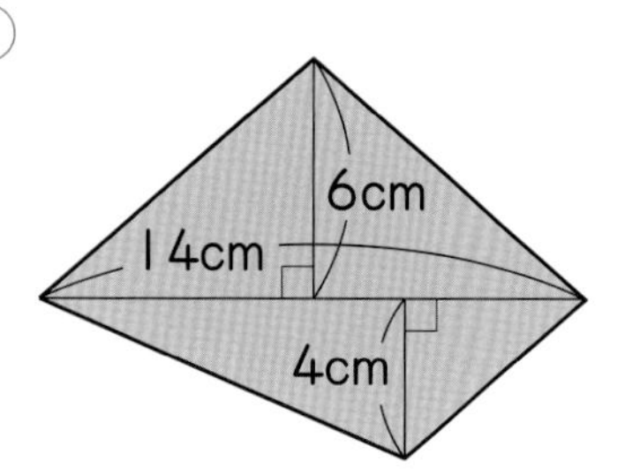


④ Parallelogram

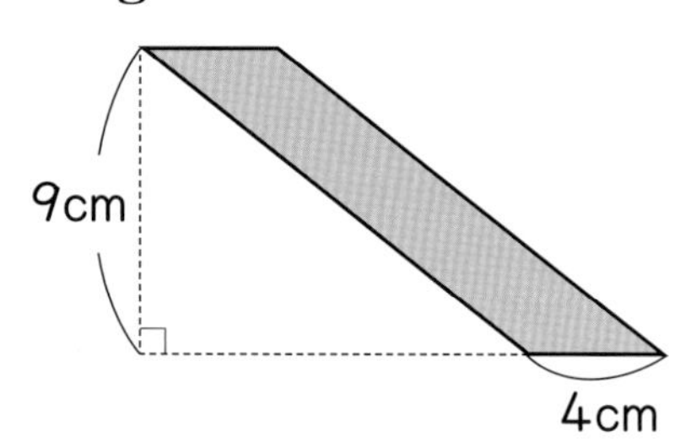


⑤ Rhombus

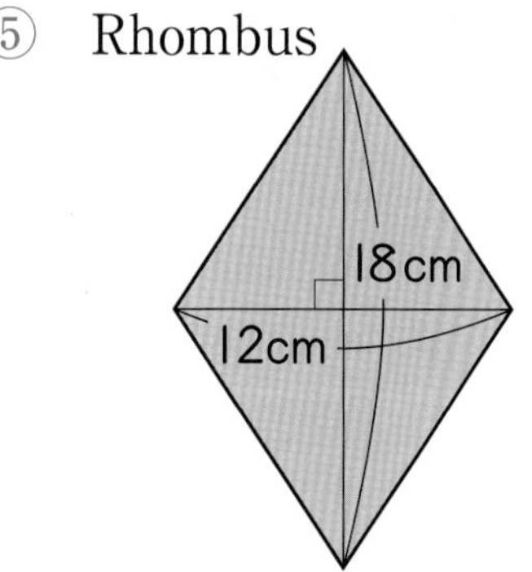


⑥ Trapezoid

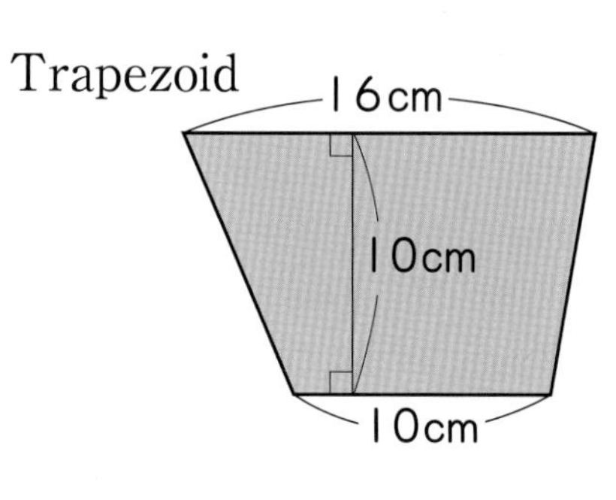


Regular polygon and circle

## 15 The angle around the center of a circle was divided into 8 equal parts to draw a regular octagon. Let's answer the following questions.

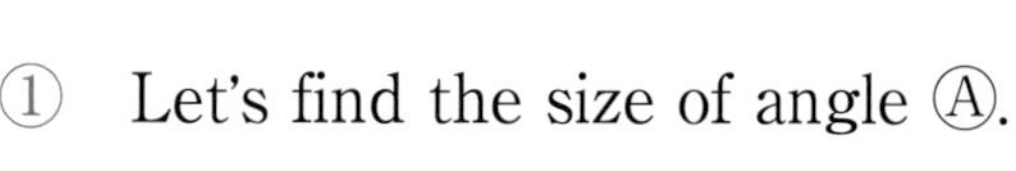

① Let's find the size of angle Ⓐ.

② Let's find the size of angle Ⓑ.

③ Let's find the size of angle Ⓒ that is one angle of a regular octagon.

Regular polygon and circle

**16** Let's find the circumference of the following circles.

① Circle with a diameter of 4 cm. ② Circle with a radius of 17.5 m.

Volume

**17** Let's find the volume of the following figures.

①
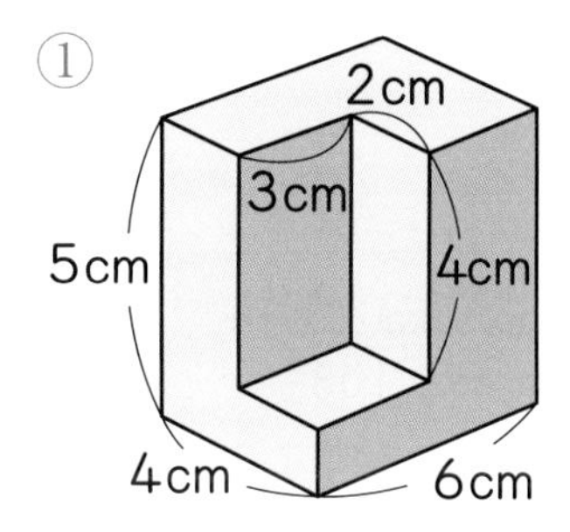


②
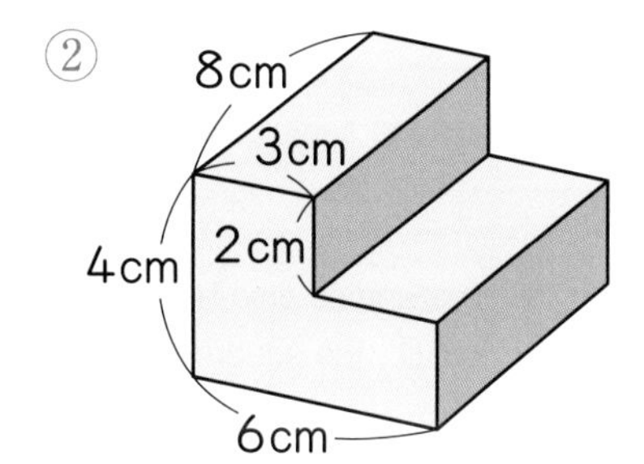


Solids

**18** Let's draw the nets of the following solids.

①
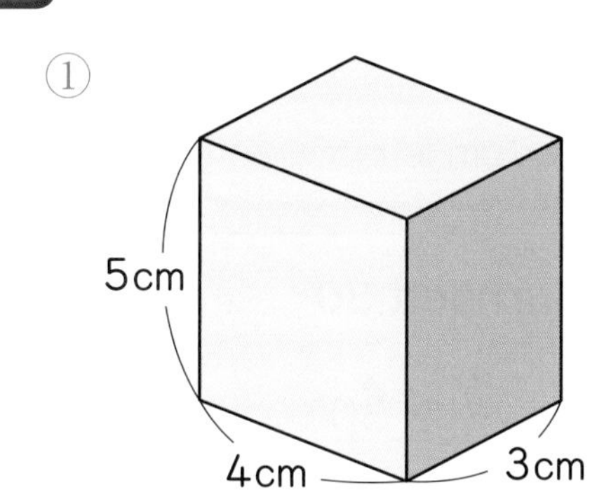


②
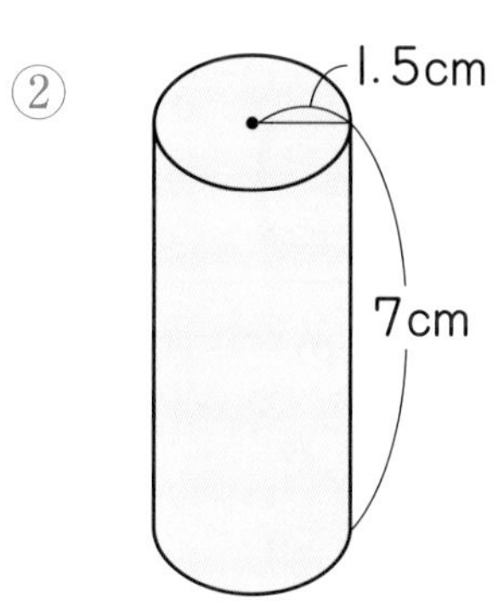


Various graphs

**19** The circle graph shown on the right shows the ratio of the number of books per type based on 160 books in a book shelf. How many Story, Biography, and Manga books are there respectively?

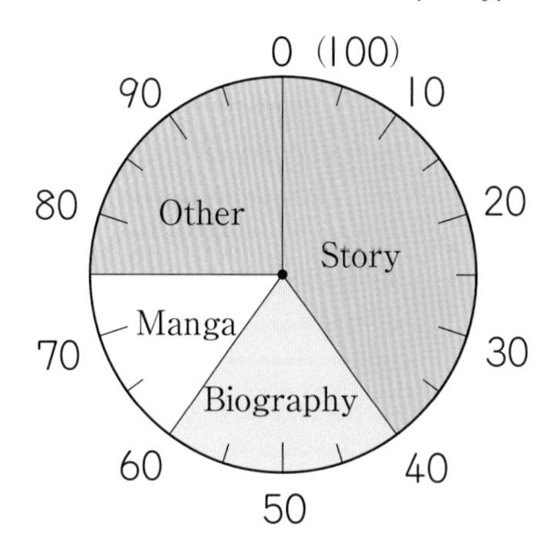

Let's produce the algorithm.

# Computational thinking

01504

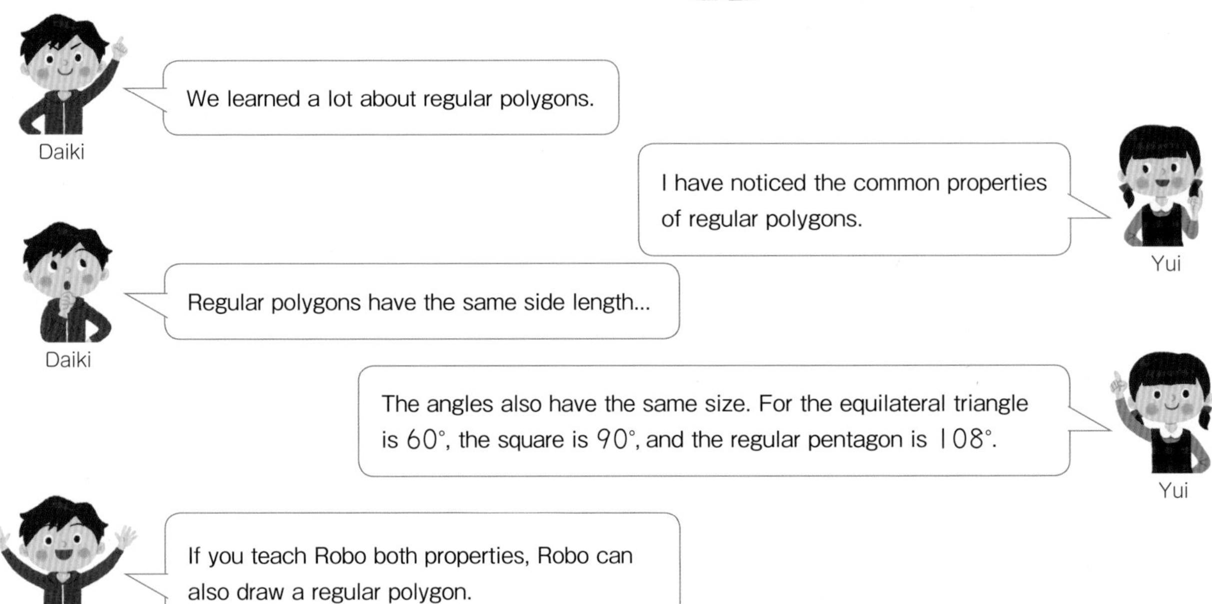

We learned a lot about regular polygons.
Daiki
I have noticed the common properties of regular polygons.
Yui
Regular polygons have the same side length...
Daiki
The angles also have the same size. For the equilateral triangle is 60°, the square is 90°, and the regular pentagon is 108°.
Yui
If you teach Robo both properties, Robo can also draw a regular polygon.
Daiki

From the above discussion, the children considered the following two instructions:

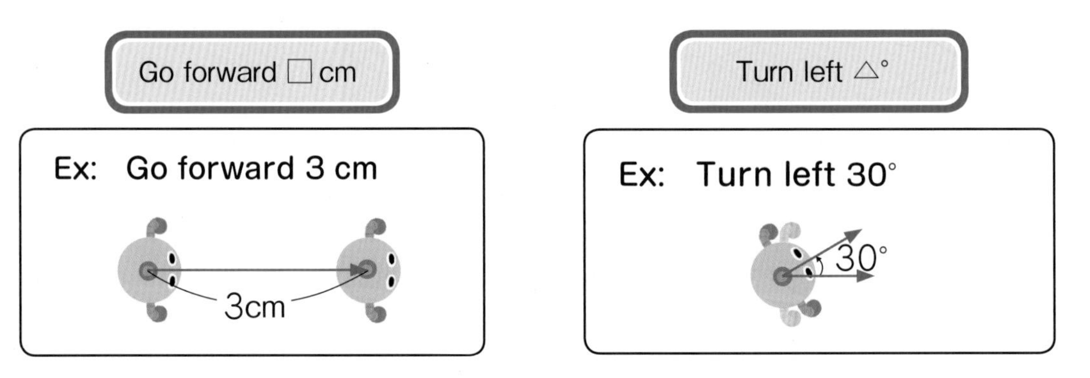


① Using the above two instructions, let's try to draw a square with a side of 4 cm with the help of Robo.

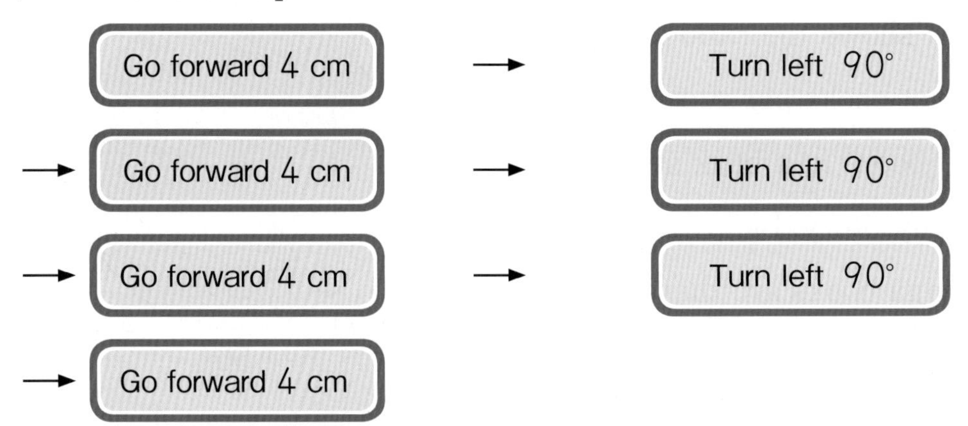

② According to the instructions in step ①, let's draw a square with a side of 4 cm in the grid below as if you were Robo.

③ Let's try to draw a regular octagon. What kind of instructions should you give to Robo?

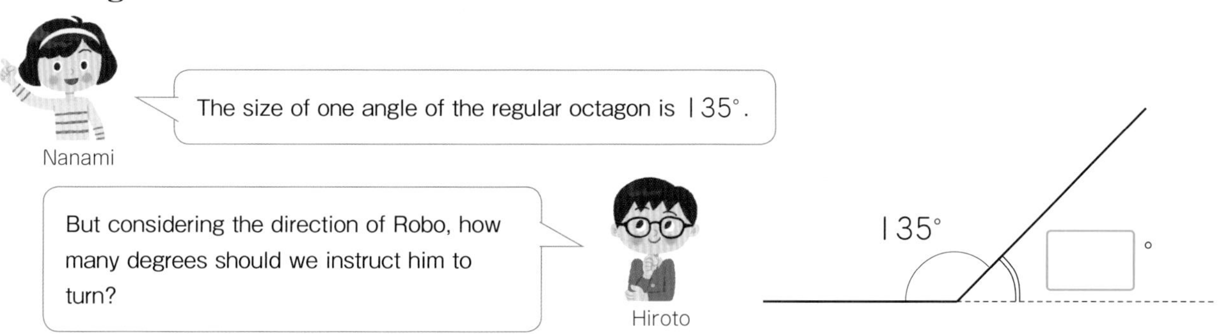

The size of one angle of the regular octagon is 135°.
Nanami
But considering the direction of Robo, how many degrees should we instruct him to turn?
Hiroto
135°
°


④ Following the instructions in step ③, let's draw a regular octagon with a side of 4 cm in the grid below as if you were Robo.

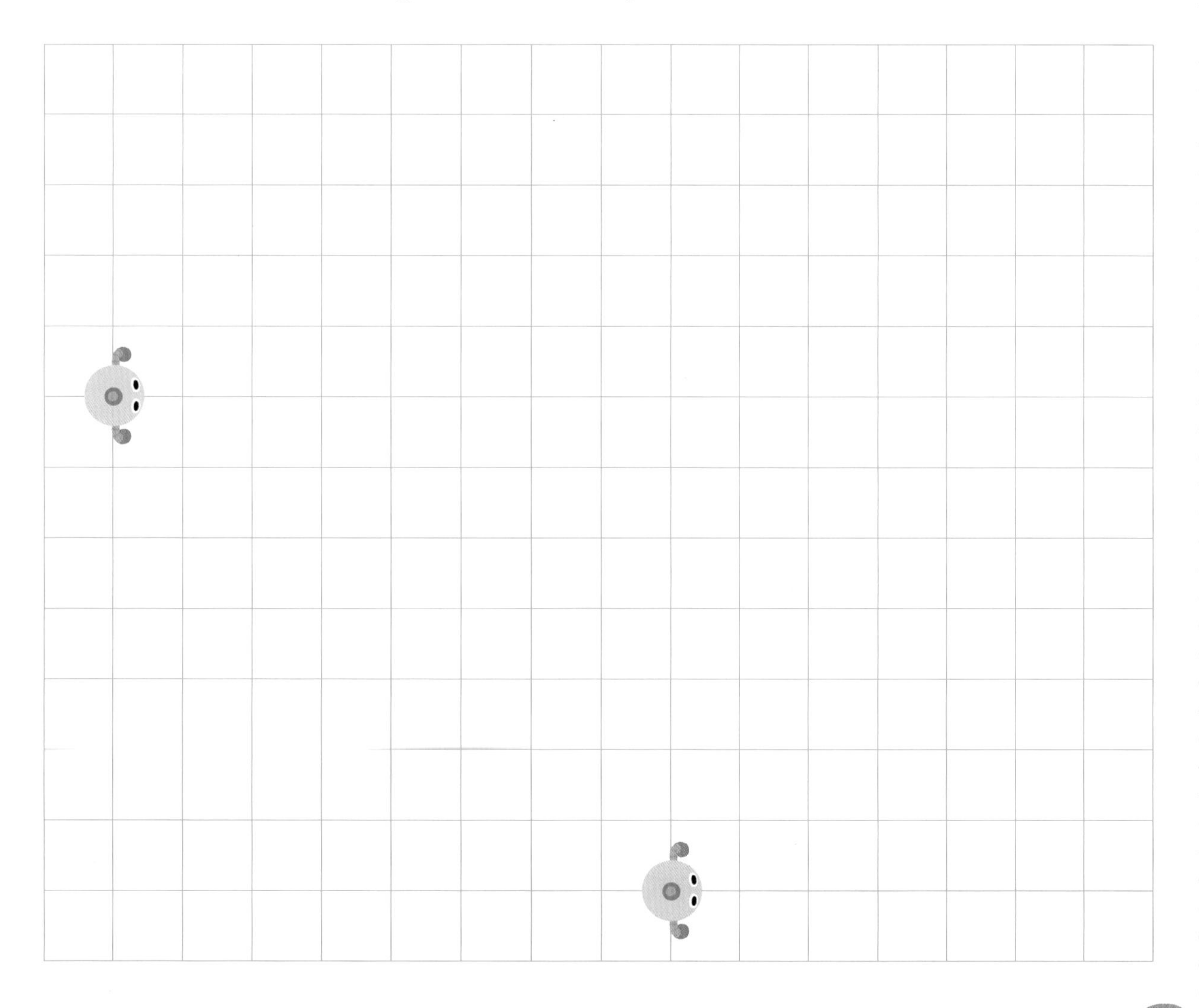

Utilize math for our life

# Let's create a tour guide book for school trip.

6th graders had a school trip.

It seems there are many destinations in Tokyo, Kyoto, and Hiroshima.

Where do they often go in elementary school?

Let's explore the situation of each place.

## Tokyo Course

Nanami

Tokyo has the largest population in Japan. Where is the population concentrated?

You can't know without exploring the population density.

Hiroto

**? Let's explore the population of each location, and represent it with a graph.**

Daiki

Is it better to represent it with a bar graph? Or are there any other graphs that are easy to understand?

Tokyo is divided into wards, cities, county and islands.

**Area and places with the most and least population in Tokyo**

| | | Place name | Population (people) | Area (km$^2$) |
|---|---|---|---|---|
| Ward | most | Setagaya | 921720 | 58 |
| | least | Chiyoda | 61218 | 12 |
| City | most | Hachioji | 577910 | 186 |
| | least | Hamura | 55225 | 10 |
| County | most | Mizuho | 33102 | 17 |
| | least | Hinohara | 2096 | 105 |
| Island | most | Oshima | 7580 | 91 |
| | least | Aogashima | 174 | 6 |

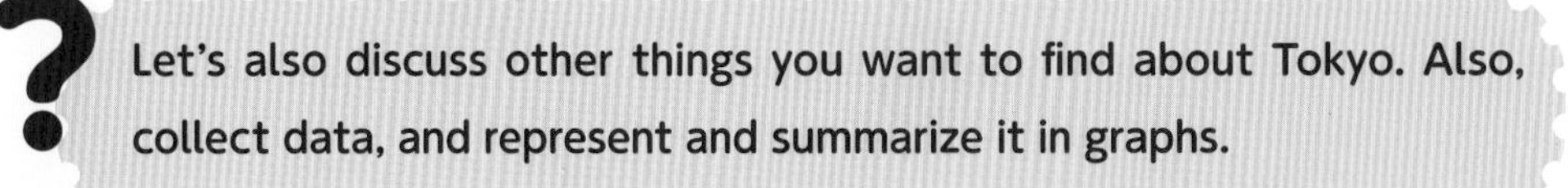

**? Let's also discuss other things you want to find about Tokyo. Also, collect data, and represent and summarize it in graphs.**

# Kyoto Course

How many elementary school students go on a school trip to Kyoto?

How has it changed from the visit in 2007 and 2017?

**Let's look at the following graphs and table, and summarize what you understand.**

**Number of elementary school students on trips to Kyoto**

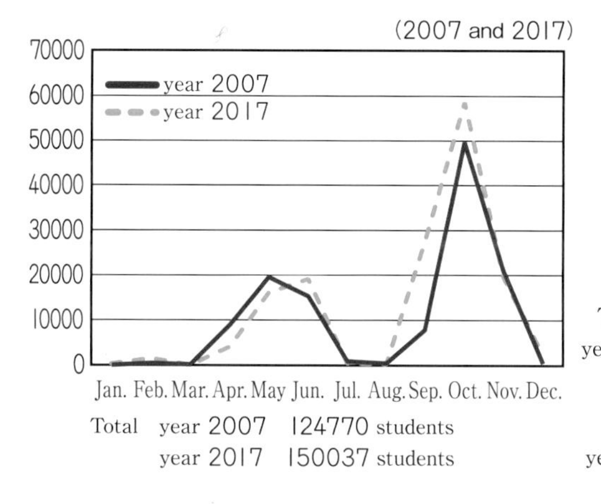


**Ratio of number of children on trip to Kyoto by departure place** (%)

| | Hokkaido | Tohoku | Kanto | Chubu | Kinki | Chugoku | Shikoku | Kyushu, Okinawa | Total |
|---|---|---|---|---|---|---|---|---|---|
| 2007 | 0 | 0.4 | 3.4 | 55.3 | 18.9 | 15.1 | 6.3 | 0.6 | 100 |
| 2017 | 0 | 0 | 2.8 | 67.2 | 13.6 | 9.4 | 5.7 | 1.3 | 100 |

**Ratio of number of children on trip to Kyoto by departure place**

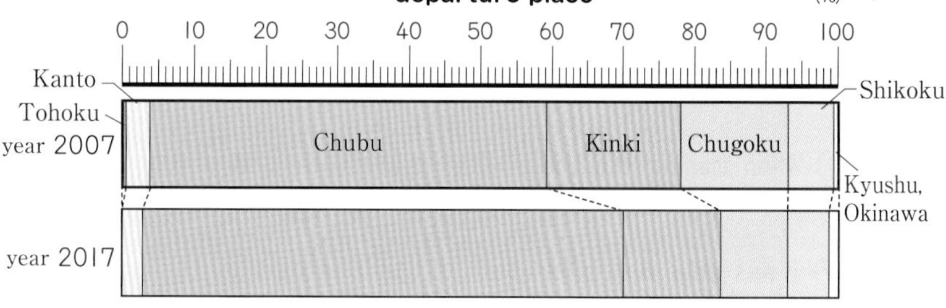


**Let's also discuss other things you want to find about Kyoto. Also, collect data, and represent and summarize it in graphs.**

# Hiroshima Course

Many people go to Hiroshima Peace Memorial Museum on school trip.

The number of foreign visitors is increasing every year.

**Number of foreign visitors to Hiroshima Peace Memorial Museum**

| Year | 2012 | 2013 | 2014 | 2015 | 2016 |
|---|---|---|---|---|---|
| Number of people | 150 thousand | 200 thousand | 230 thousand | 340 thousand | 370 thousand |

**Let's represent the above table with a line graph. Let's look at the line graph and predict the number of foreign visitors in 2019.**

# Let's create a tour guide book for school trip.

## 1. Toward learning competency.

| | Strongly agree | Agree | Don't agree |
|---|---|---|---|
| ① It was fun working on it. | | | |
| ② The learning contents were helpful. | | | |
| ③ Concentrated our efforts to make something better. | | | |

## 2. Thinking, deciding and representing competency.

| | Definitely did | So so | I didn't |
|---|---|---|---|
| ① When making the tour guide book, I was able to discover where to use the mathematical knowledge. | | | |
| ② I was able to confirm how to use mathematical knowledge and whether the numerical values were correct. | | | |
| ③ For the tour guide book, I was able to represent my mathematical knowledge in text, pictures, figures, and tables. | | | |

## 3. What I know and can do.

| | Definitely did | So so | I didn't |
|---|---|---|---|
| ① I was able to make a better tour guide book. | | | |
| ② Making a guide book has deepened on my understanding of mathematical knowledge. | | | |

## 4. Encouragement for myself.

| | Strongly agree |
|---|---|
| ① I think that I'm doing my best. | |

Give yourself a compliment since you have worked so hard.

Let's collect more data, change topic, and try to summarize what you were not able to accomplish and keep doing your best on what you were able to fulfill.

# Supplementary Problems

## 11 Addition and Subtraction of Fractions

p.4~p.19

**1** Let's reduce the following fractions.

① $\frac{10}{15}$　② $\frac{8}{24}$

③ $\frac{32}{40}$　④ $\frac{7}{42}$

⑤ $\frac{10}{45}$　⑥ $2\frac{30}{36}$

**2** Let's change the following pair of fractions into fractions with common denominators and write inequality signs in the □.

① $\frac{2}{3}$ □ $\frac{1}{2}$　② $\frac{3}{4}$ □ $\frac{7}{9}$

③ $\frac{2}{5}$ □ $\frac{3}{10}$　④ $\frac{5}{7}$ □ $\frac{11}{14}$

⑤ $\frac{5}{8}$ □ $\frac{7}{12}$　⑥ $\frac{8}{15}$ □ $\frac{7}{10}$

**3** Let's write the following three fractions in ascending order: $\frac{8}{15}$, $\frac{5}{6}$, $\frac{7}{10}$.

**4** Let's solve the following calculations.

① $\frac{1}{5}+\frac{3}{7}$　② $\frac{11}{16}+\frac{7}{8}$

③ $\frac{1}{4}+\frac{3}{10}$　④ $\frac{4}{9}+\frac{1}{12}$

⑤ $\frac{5}{12}+\frac{5}{6}$　⑥ $\frac{1}{6}+\frac{4}{15}$

⑦ $1\frac{1}{2}+1\frac{2}{5}$　⑧ $1\frac{2}{21}+2\frac{4}{7}$

**5** Let's solve the following calculations.

① $3\frac{2}{3}+4\frac{4}{5}$　② $\frac{3}{4}+3\frac{3}{5}$

③ $2\frac{7}{18}+\frac{7}{9}$　④ $2\frac{5}{12}+3\frac{3}{4}$

⑤ $1\frac{7}{10}+2\frac{5}{6}$　⑥ $2\frac{13}{15}+5\frac{3}{10}$

**6** Let's solve the following calculations.

① $\frac{2}{3}-\frac{2}{5}$　② $\frac{11}{15}-\frac{4}{9}$

③ $\frac{5}{7}-\frac{3}{14}$　④ $\frac{7}{12}-\frac{9}{20}$

**7** Let's solve the following calculations.

① $\frac{8}{5}-\frac{2}{3}$　② $\frac{5}{4}-\frac{8}{7}$

③ $4\frac{2}{5}-3\frac{1}{9}$　④ $8\frac{5}{8}-6\frac{5}{12}$

⑤ $3\frac{1}{7}-\frac{9}{14}$　⑥ $6\frac{1}{6}-2\frac{7}{15}$

**8** Let's solve the following calculations.

① $\frac{3}{4}+\frac{1}{2}-\frac{5}{16}$

② $\frac{11}{12}-\frac{1}{6}-\frac{3}{10}$

**9** There are $\frac{7}{8}$ L and $1\frac{5}{12}$ L of juice. Let's answer the following questions.

① How many liters of juice are there altogether?

② How many liters is the difference between the two amounts of juice?

## ⑫ Fractions, Decimal Numbers, and Whole Numbers

p.20~p.29

**1** If you divide 5 L of juice into 6 equal parts, how many liters is the amount in one part?

**2** Let's represent the following quotients as fractions.

① $2 \div 5$　② $3 \div 7$

③ $4 \div 9$　④ $5 \div 3$

**3** Let's answer about the number of sheets of colored paper shown on the right.

**Sheets of colored paper**

| Color | Number of sheets of paper |
|---|---|
| Red | 9 |
| Blue | 7 |
| Yellow | 10 |

① How many times of the number of sheets of red paper is the number of sheets of blue paper?

② How many times of the number of sheets of red paper is the number of sheets of yellow paper?

**4** The distance from the house to the school is 300 m and the distance from the house to the station is 700 m. Let's answer the following questions.

① How many times of the distance from the house to the school is the distance from the house to the station?

② How many times of the distance from the house to the station is the distance from the house to the school?

**5** Let's represent the following fractions as decimal numbers.

① $\frac{9}{10}$　② $\frac{33}{100}$

③ $\frac{4}{5}$　④ $1\frac{3}{4}$

**6** Let's represent the following fractions as whole numbers.

① $\frac{8}{2}$　② $\frac{18}{9}$

**7** Let's represent the following decimal numbers as fractions.

① 0.7　② 0.27

③ 0.4　④ 0.32

⑤ 2.5　⑥ 1.15

**8** Let's write the numbers that apply in the following □.

① $8 = \frac{\square}{1}$　② $3 = \frac{\square}{5}$

③ $6 = \frac{\square}{2}$　④ $4 = \frac{\square}{4}$

**9** Let's write the inequality signs that apply in the following □.

① $\frac{2}{5}$ □ 0.5　② 0.7 □ $\frac{3}{4}$

③ $\frac{5}{6}$ □ 0.9　④ 0.6 □ $\frac{7}{12}$

⑤ 1.2 □ $\frac{8}{7}$　⑥ $\frac{25}{8}$ □ 3.2

**10** Let's represent the following numbers by using ↑ on the number line below.

0.3　$\frac{4}{5}$　$1\frac{5}{10}$　1.2　$\frac{18}{20}$

0　　　1

## 13 Ratio (1)

p.30~p.41

**1** Jun's Elementary School has a total number of 550 students. Among them, 99 students are from 5th grade. Let's find, as a decimal number, the ratio of students from 5th grade based on the total number of students.

**2** Let's change the following ratios from decimal numbers to percentages and from percentages to decimal numbers.

① 0.85　　② 0.7
③ 0.924　　④ 0.036
⑤ 38%　　⑥ 40%
⑦ 5%　　⑧ 17.2%

**3** Let's change the following ratios from decimal numbers to 歩合 (buai) and from 歩合 (buai) to decimal numbers.

① 0.9　　② 0.68
③ 0.521　　④ 0.207
⑤ 3割5分5厘 (3wari 5bu 5rin)
⑥ 8割(8wari)
⑦ 7割6分(7wari 6bu)
⑧ 4分7厘(4bu 7rin)

**4** A baseball team won 16 games out of 25 games. What percentage of the total number of games is the ratio of games won?

**5** I bought a handkerchief with a fixed price of 600 yen for 480 yen. How many 割 (wari) is the ratio of the paid price based on the fixed price?

# 14 Area of Figures

p.42~p.63

**1** Let's find the area of the following parallelograms.

① ②

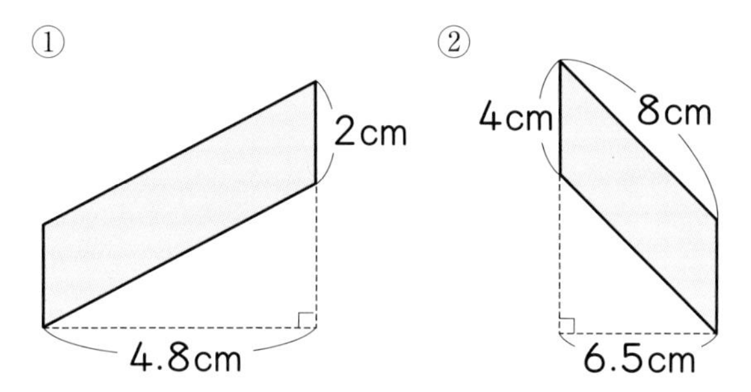


**2** Let's find the area of the following triangles.

① ②

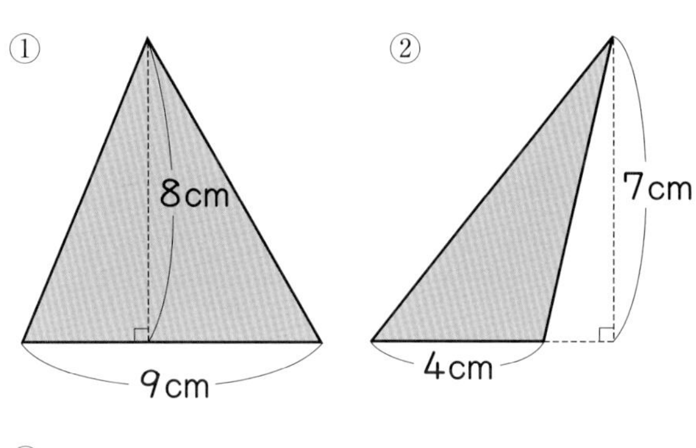


③

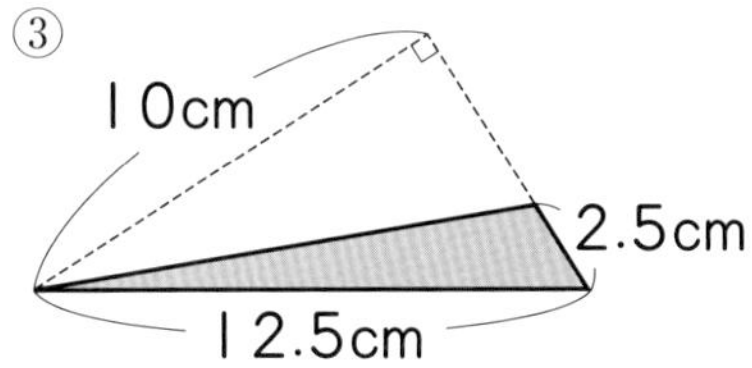


**3** Let's find the height of the triangle shown below when the base is side AB.

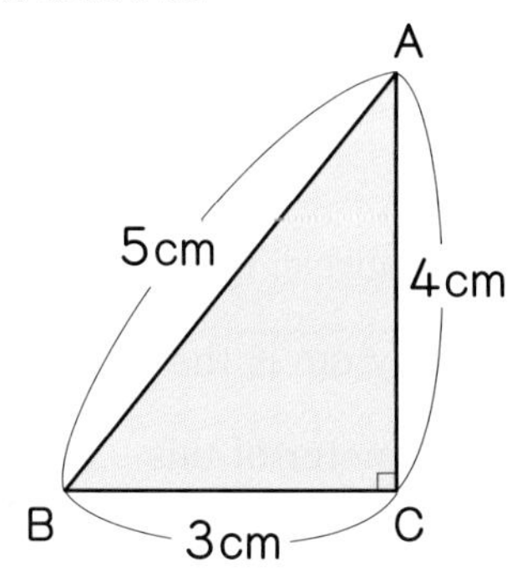


**4** Let's find the area of the following trapezoids.

① ②

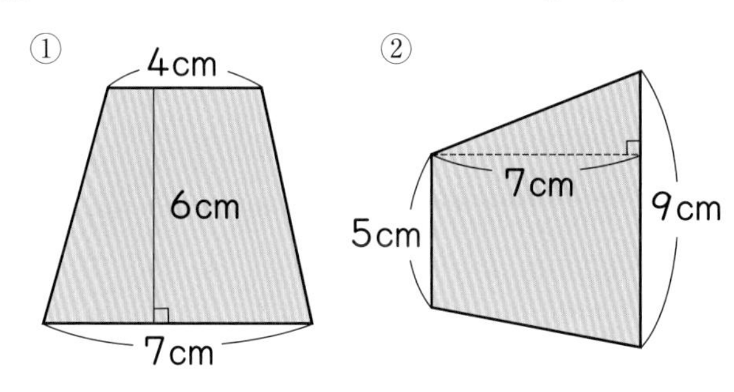


**5** Let's find the area of the following rhombuses.

① ②

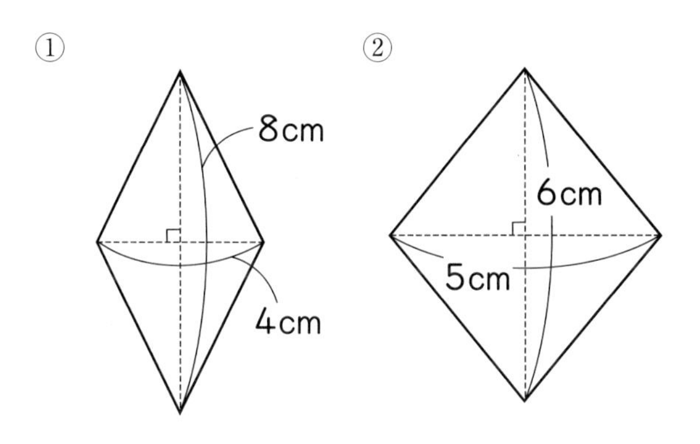


**6** Let's find the area of the following figures.

①

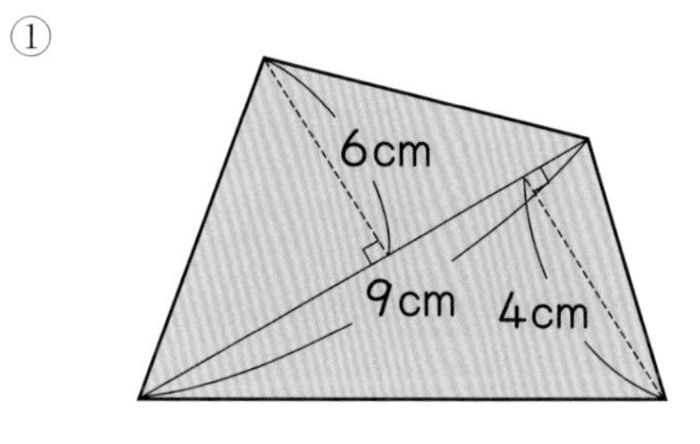


②

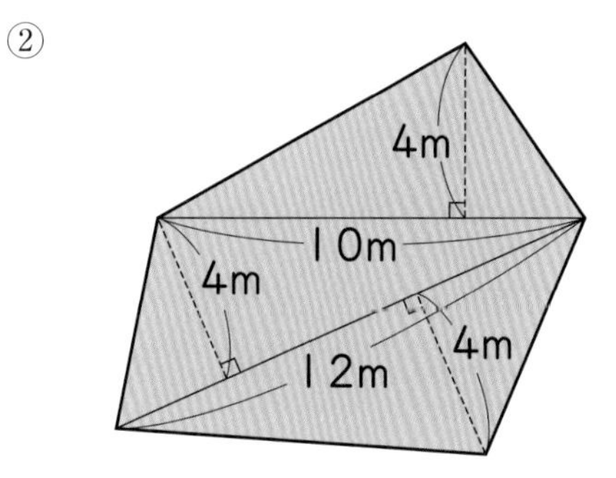

# ⑮ Regular Polygon and Circle

p.68~p.82

**1** The figure shown on the right is a regular hexagon. How many degrees is the size of angles ⓐ and ⓑ?

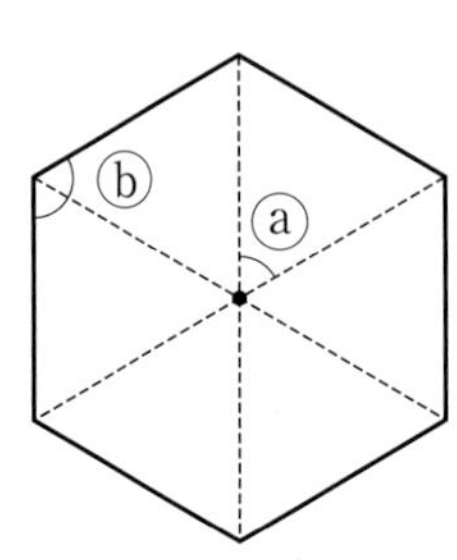


**2** The figure shown on the right is a regular octagon. How many degrees is the size of angles ⓐ and ⓑ?

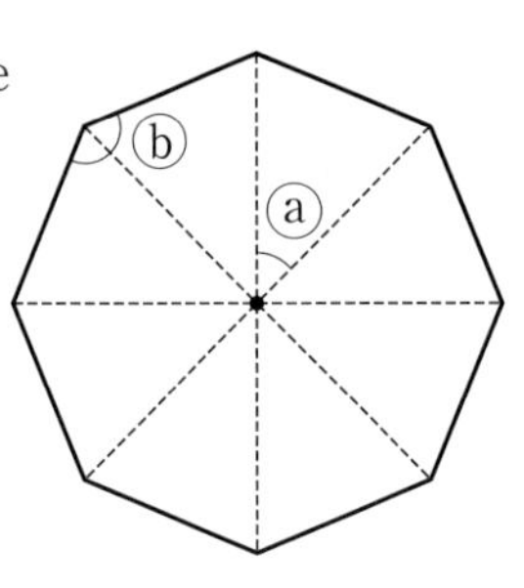


**3** Let's draw a regular pentagon by using the following circle.

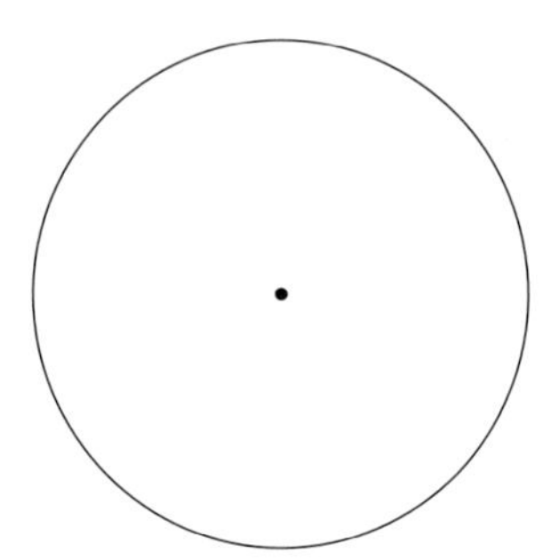

**4** What is the name of the number that can be found with circumference ÷ diameter?

**5** Let's find the circumference.

①

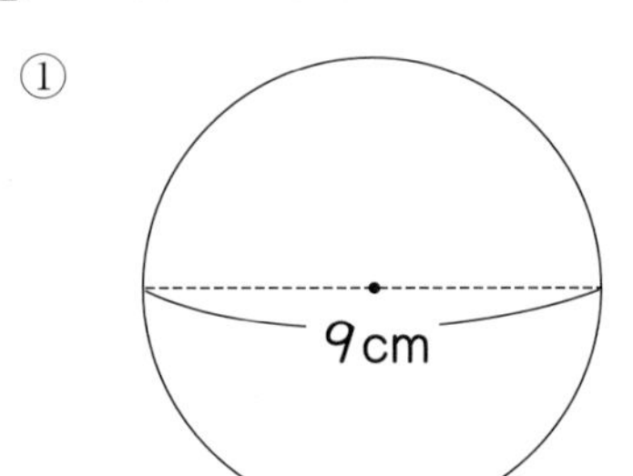


②

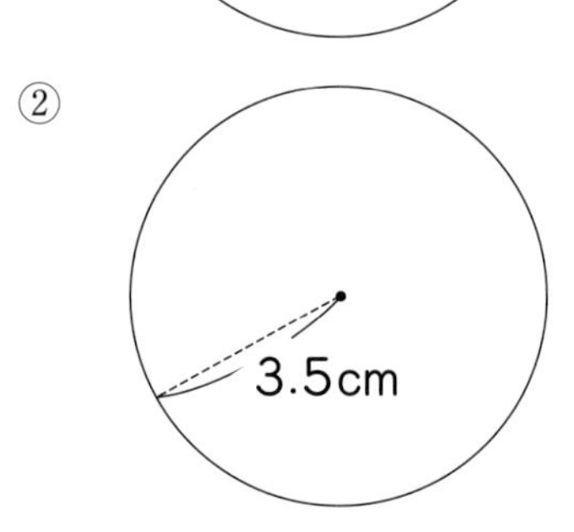


**6** Let's find the diameter of the circles with the following lengths as circumference.

① 18.84 cm ② 47.1 cm

③ 43.96 cm ④ 78.5 cm

**7** Let's find the diameter of the circles with the following lengths as circumference. Let's find the nearest whole round number by rounding off the tenths place.

① 234 cm ② 658 cm

**8** The following figure was made by combining the diameters of three semicircles on straight line AB. The points on straight line AB are the centers of each semicircle. How many cm is the perimeter of this figure?

# 16 Volume

p.83~p.101

**1** How many $cm^3$ is the volume of the following cuboid and cube?

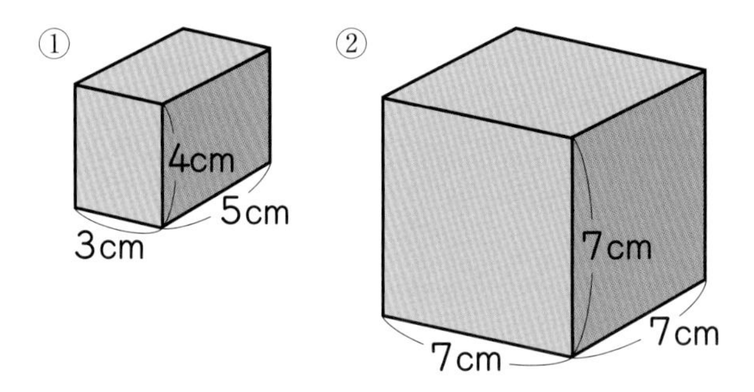


**2** How many $cm^3$ is the volume of the cuboid that can be assembled with the following net?

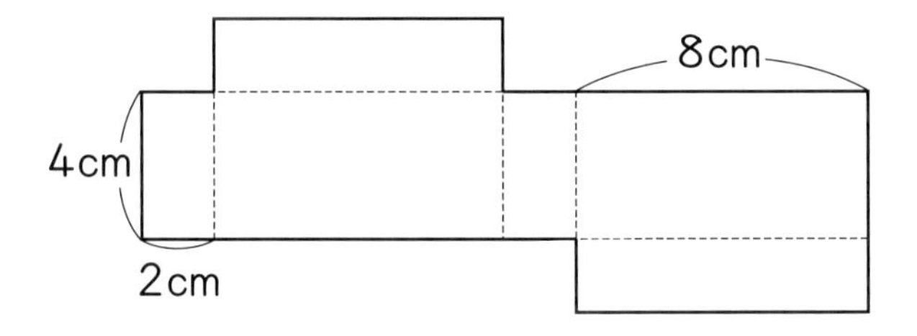


**3** How many $m^3$ is the volume of the following cuboid? Also, how many $cm^3$?

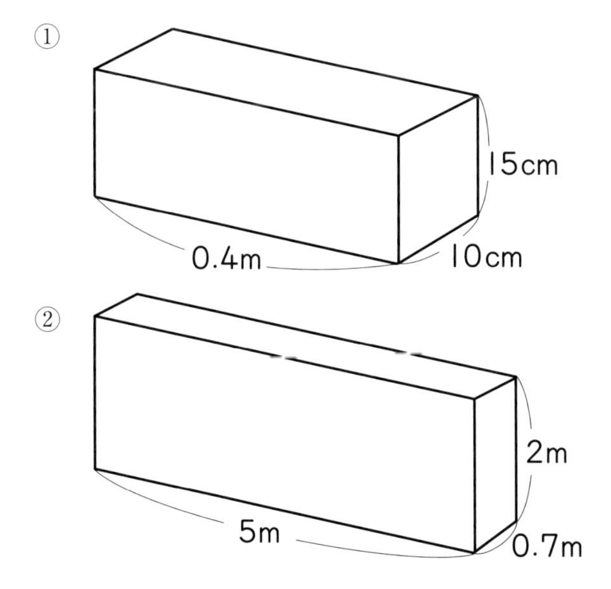


**4** Let's write the numbers that apply in the following ☐.

① $2\ m^3 =$ ☐ $cm^3$

② $5000\ L =$ ☐ $m^3$

③ $30\ mL =$ ☐ $cm^3$

**5** Let's find the volume of the following figures.

①

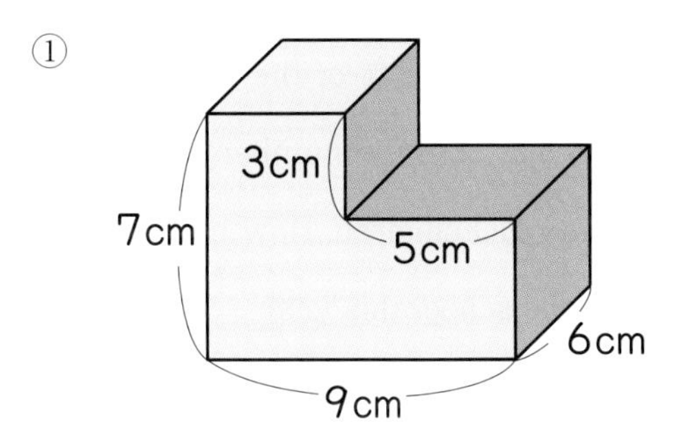


②

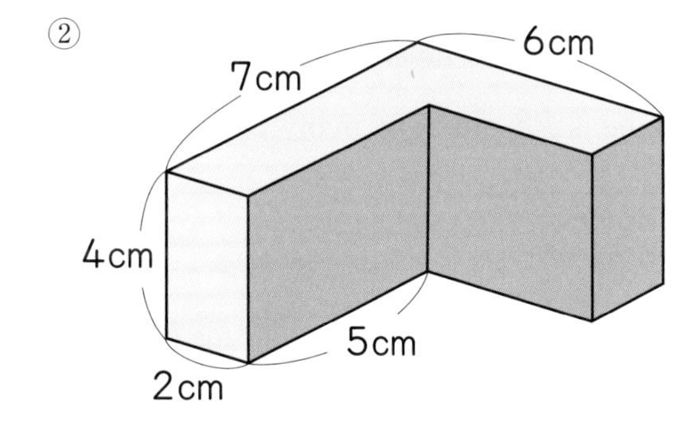


③

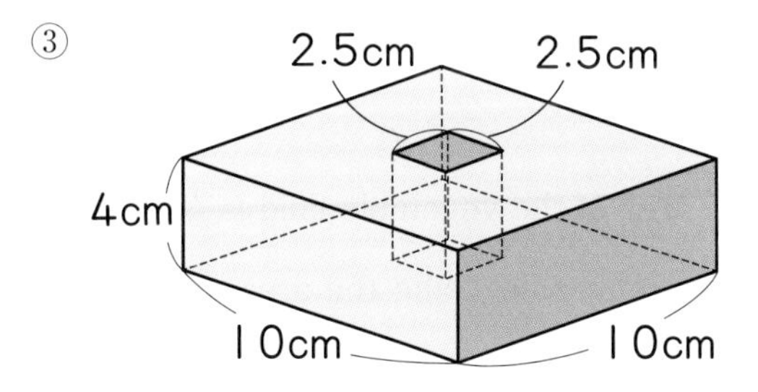


**6** There is a cuboid water tank with an inside length, inside width, and depth of 20 cm, 30 cm, and 15 cm respectively. How many liters of water will be poured in if the tank is filled completely?

## ⑰ Ratio (2)

p.102~p.112

**1** A train car with a capacity of 120 people has 132 passengers. What is the percentage of the number of passengers compared to the capacity?

**2** The number of children in an elementary school this year is 108 % of the number of children last year, and the number of children last year was 600 children. How many children are there in the school this year?

**3** The length of the red tape is 125 % of the length of the blue tape, and the length of the red tape is 28 cm. How many cm is the length of the blue tape?

**4** At a store, goods were bought for 800 yen and will be sold with an added 15 % profit. How many yen should the selling price be?

**5** The area of the vegetable field is 40 % larger than the flower field, and the area of the flower field is 15 $m^2$. How many $m^2$ is the area of the vegetable field?

**6** Sweaters are sold for 1800 yen. When this price is 2割 (2 wari) more expensive than the price of a shirt, how many yen is the price of the shirt?

## ⑱ Various Graphs

p.113~p.121

**1** The following strip graph shows the ratio of components included in soybeans.

**Components included in soybean**

| Protein 34% | Starch 30% | Fat | Water 12% | |
|---|---|---|---|---|

Other 4%

① What percentage is the ratio of fat?

② How many grams of protein and starch are included in 300 g of soybeans?

**2** The table below shows the ratio of land use in Nakagawa Town. Let's answer the following questions.

**Land use in Nakagawa Town** (%)

| Manufacturing | Farming | Commerce | Housing | Other |
|---|---|---|---|---|
| 35 | 28 | 15 | 12 | 10 |

① Let's represent this in a circle graph.

**Land use in Nakagawa Town**

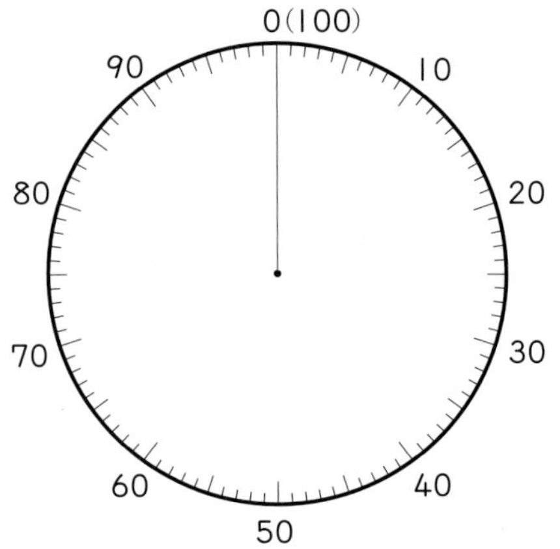


② When the farm area in Nakagawa Town is 8.4 $km^2$, how many $km^2$ is the area of the whole town?

# 19 Solids

p.122~p.131

**1** Let's answer about the following three solids.

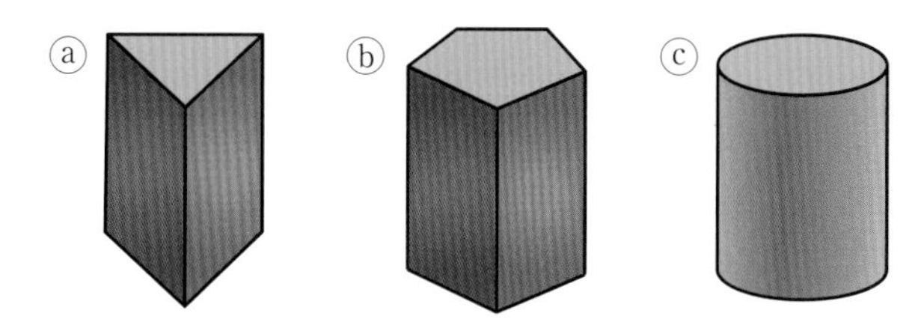


① What kind of solid are they?

② Which of the solids are covered only by planes?

③ Which of the solids is covered by planes and curved faces?

**2** Let's write the words that apply in the following ☐.

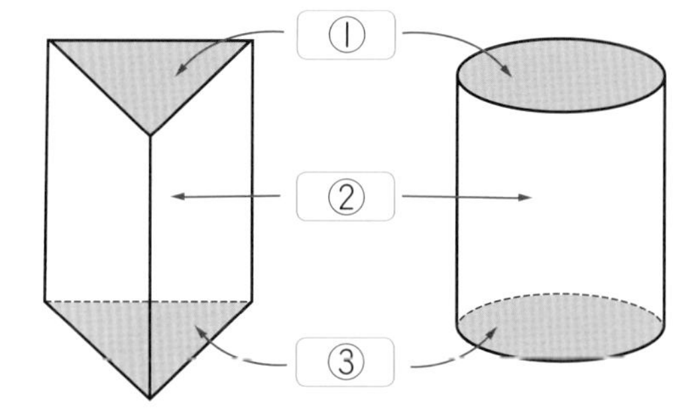


**3** Let's summarize about triangular prisms, quadrangular prisms, and pentagonal prisms in the following table.

| | Triangular prism | Quadrangular prism | Pentagonal prism |
|---|---|---|---|
| Number of faces | 5 | | |
| Number of vertices | | 8 | |
| Number of edges | | | 15 |

**4** Let's answer about the triangular prism shown on the right. The bases are equilateral triangles.

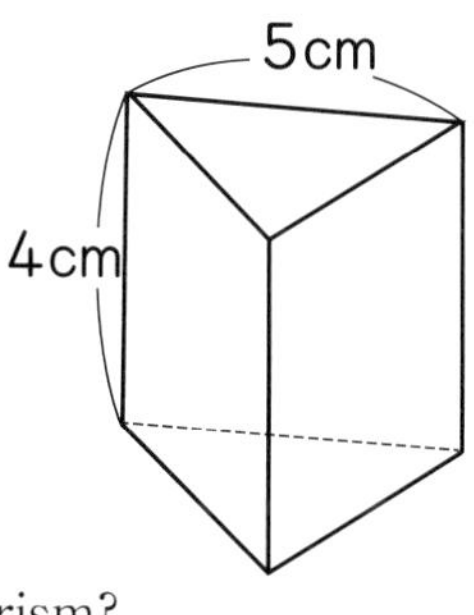


① How many cm is the height of this triangular prism?

② Let's draw the net of this triangular prism.

**5** Let's answer about the net of a cylinder shown below.

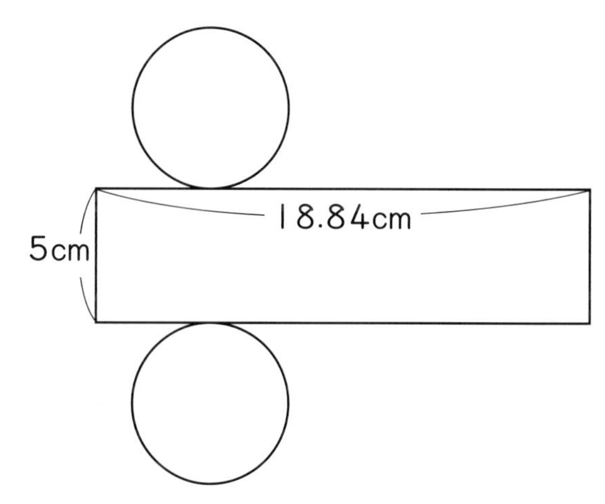


① How many cm is the height of this cylinder?

② How many cm is the diameter of a circle in the base of this cylinder?

**6** Let's draw the net of the following cylinder.

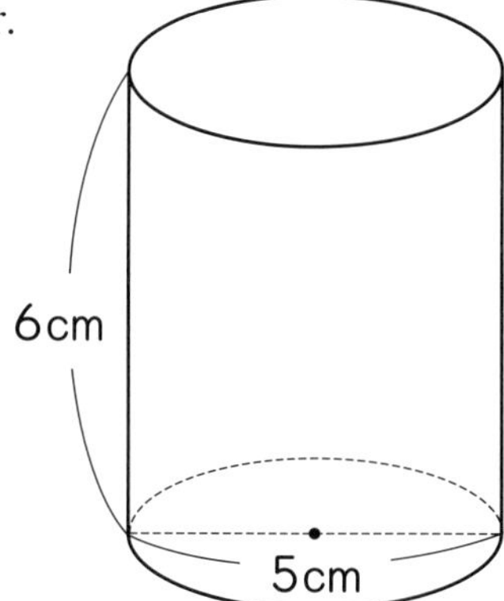

## 11 Addition and Subtraction of Fractions

**1** ① $\frac{2}{3}$ ② $\frac{1}{3}$ ③ $\frac{4}{5}$ ④ $\frac{1}{6}$ ⑤ $\frac{2}{9}$ ⑥ $2\frac{5}{6}$

**2** ① $>$ ② $<$ ③ $>$ ④ $<$ ⑤ $>$ ⑥ $<$

**3** $\frac{8}{15}$, $\frac{7}{10}$, $\frac{5}{6}$

**4** ① $\frac{22}{35}$ ② $1\frac{9}{16}\left(\frac{25}{16}\right)$ ③ $\frac{11}{20}$ ④ $\frac{19}{36}$

⑤ $1\frac{1}{4}\left(\frac{5}{4}\right)$ ⑥ $\frac{13}{30}$ ⑦ $2\frac{9}{10}$ ⑧ $3\frac{2}{3}$

**5** ① $8\frac{7}{15}$ ② $4\frac{7}{20}$ ③ $3\frac{1}{6}$ ④ $6\frac{1}{6}$

⑤ $4\frac{8}{15}$ ⑥ $8\frac{1}{6}$

**6** ① $\frac{4}{15}$ ② $\frac{13}{45}$ ③ $\frac{1}{2}$ ④ $\frac{2}{15}$

**7** ① $\frac{14}{15}$ ② $\frac{3}{28}$ ③ $1\frac{13}{45}$ ④ $2\frac{5}{24}$

⑤ $2\frac{1}{2}$ ⑥ $3\frac{7}{10}$

**8** ① $\frac{15}{16}$ ② $\frac{9}{20}$

**9** ① $\frac{7}{8} + 1\frac{5}{12} = 2\frac{7}{24}$ $2\frac{7}{24}$ L

② $1\frac{5}{12} - \frac{7}{8} = \frac{13}{24}$ $\frac{13}{24}$ L

## 12 Fractions, Decimal Numbers, and Whole Numbers

**1** $5 \div 6 = \frac{5}{6}$ $\frac{5}{6}$ L

**2** ① $\frac{2}{5}$ ② $\frac{3}{7}$ ③ $\frac{4}{9}$ ④ $\frac{5}{3}$

**3** ① $7 \div 9 = \frac{7}{9}$ $\frac{7}{9}$ times

② $10 \div 9 = 1\frac{1}{9}$ $1\frac{1}{9}$ times $\left(\frac{10}{9}\text{ times}\right)$

**4** ① $700 \div 300 = 2\frac{1}{3}$ $2\frac{1}{3}$ times $\left(\frac{7}{3}\text{ times}\right)$

② $300 \div 700 = \frac{3}{7}$ $\frac{3}{7}$ times

**5** ① 0.9 ② 0.33 ③ 0.8 ④ 1.75

**6** ① 4 ② 2

**7** ① $\frac{7}{10}$ ② $\frac{27}{100}$ ③ $\frac{4}{10}\left(\frac{2}{5}\right)$ ④ $\frac{32}{100}\left(\frac{8}{25}\right)$

⑤ $\frac{25}{10}\left(\frac{5}{2}\right)$ ⑥ $\frac{115}{100}\left(\frac{23}{20}\right)$

**8** ① 8 ② 15 ③ 12 ④ 16

**9** ① $<$ ② $<$ ③ $<$ ④ $>$ ⑤ $>$ ⑥ $<$

**10**

0 1

0.3 $\frac{4}{5}$ $\frac{18}{20}$ 1.2 $1\frac{5}{10}$

## 13 Ratio (1)

**1** $99 \div 550 = 0.18$ 0.18

**2** ① 85% ② 70% ③ 92.4% ④ 3.6%

⑤ 0.38 ⑥ 0.4 ⑦ 0.05 ⑧ 0.172

**3** ① 9割 (9wari) ② 6割8分 (6wari 8bu)

③ 5割2分1厘 (5wari 2bu 1rin)

④ 2割7厘 (2wari 7rin)

⑤ 0.355 ⑥ 0.8 ⑦ 0.76 ⑧ 0.047

**4** $16 \div 25 = 0.64$ 64%

**5** $480 \div 600 = 0.8$ 8割 (8wari)

## 14 Area of Figures

**1** ① 9.6 cm² ② 26 cm²

**2** ① 36 cm² ② 14 cm² ③ 12.5 cm²

**3** 2.4 cm

**4** ① 33 cm² ② 49 cm²

**5** ① 16 cm² ② 15 cm²

**6** ① 45 cm² ② 68 m²

## 15 Regular Polygon and Circle

**1** ⓐ 60° ⓑ 120°

**2** ⓐ 45° ⓑ 135°

**3**

**4** Ratio of circumference

**5** ① 28.26 cm ② 21.98 cm

**6** ① 6 cm ② 15 cm ③ 14 cm ④ 25 cm

**7** ① 75 cm ② 210 cm

**8** 31.4 cm

## ⑯ Volume

1 ① 60 cm³ ② 343 cm³

2 64 cm³

3 ① 0.006 m³, 6000 cm³ ② 7 m³, 7000000 cm³

4 ① 2000000 ② 5 ③ 30

5 ① 288 cm³ ② 88 cm³ ③ 375 cm³

6 9 L

## ⑰ Ratio (2)

1 $132 \div 120 = 1.1$ 110%

2 $600 \times 1.08 = 648$ 648 children

3 $28 \div 1.25 = 22.4$ 22.4 cm

4 $800 \times (1 + 0.15) = 920$ 920 yen

5 $15 \times (1 + 0.4) = 21$ 21 m²

6 $1800 \div (1 + 0.2) = 1500$ 1500 yen

## ⑱ Various Graphs

1 ① $100 - (34 + 30 + 12 + 4) = 20$ 20%

② Protein $300 \times 0.34 = 102$ 102 g

Starch $300 \times 0.3 = 90$ 90 g

2 ①

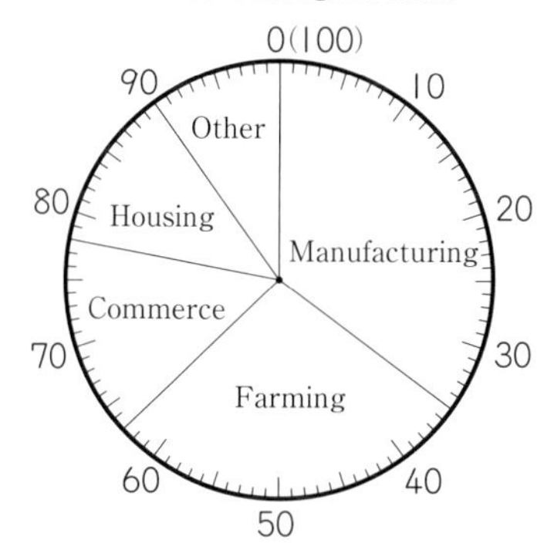


② $8.4 \div 0.28 = 30$ 30 km²

## ⑲ Solids

1 ① ⓐ triangular prism ⓑ pentagonal prism ⓒ cylinder

② ⓐ, ⓑ ③ ⓒ

2 ① base ② lateral face ③ base

3

| | Triangular prism | Quadrangular prism | Pentagonal prism |
|---|---|---|---|
| Number of faces | 5 | 6 | 7 |
| Number of vertices | 6 | 8 | 10 |
| Number of edges | 9 | 12 | 15 |

4 ① 4 cm ②

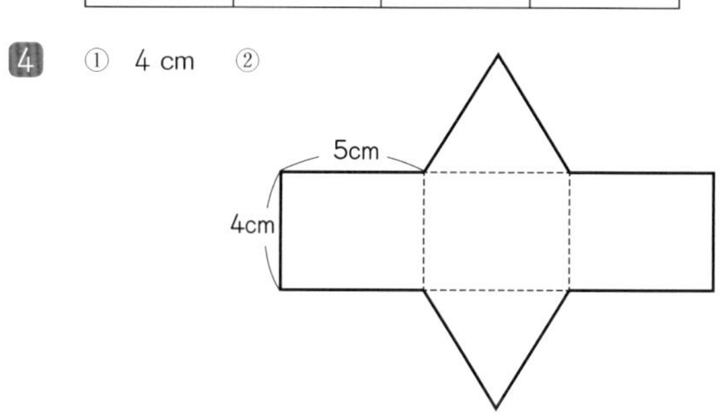


5 ① 5 cm ② 6 cm

6

## Words and symbols from this book.

# Regular Polygon and Circle

▼ will be used in page 74.

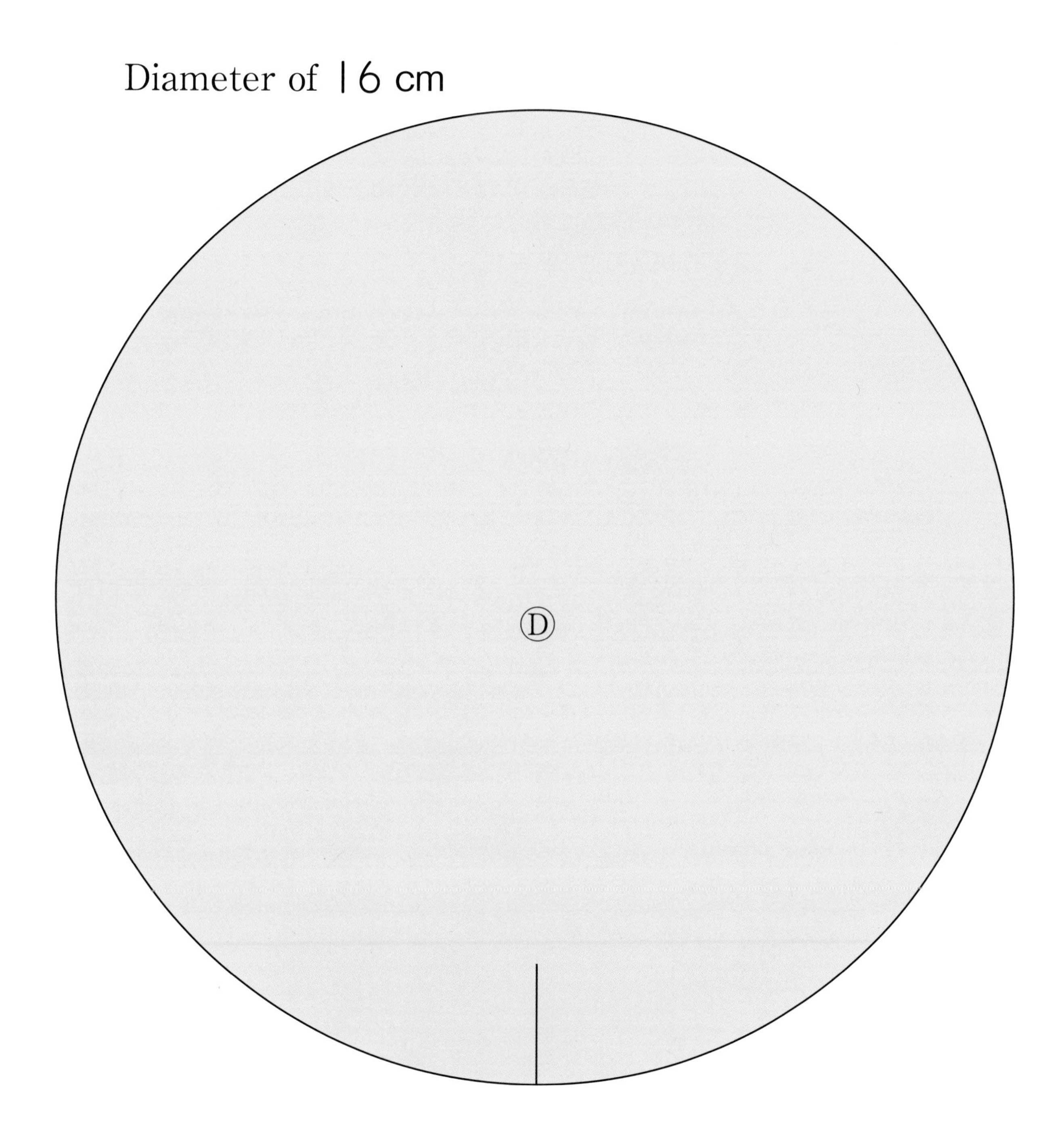

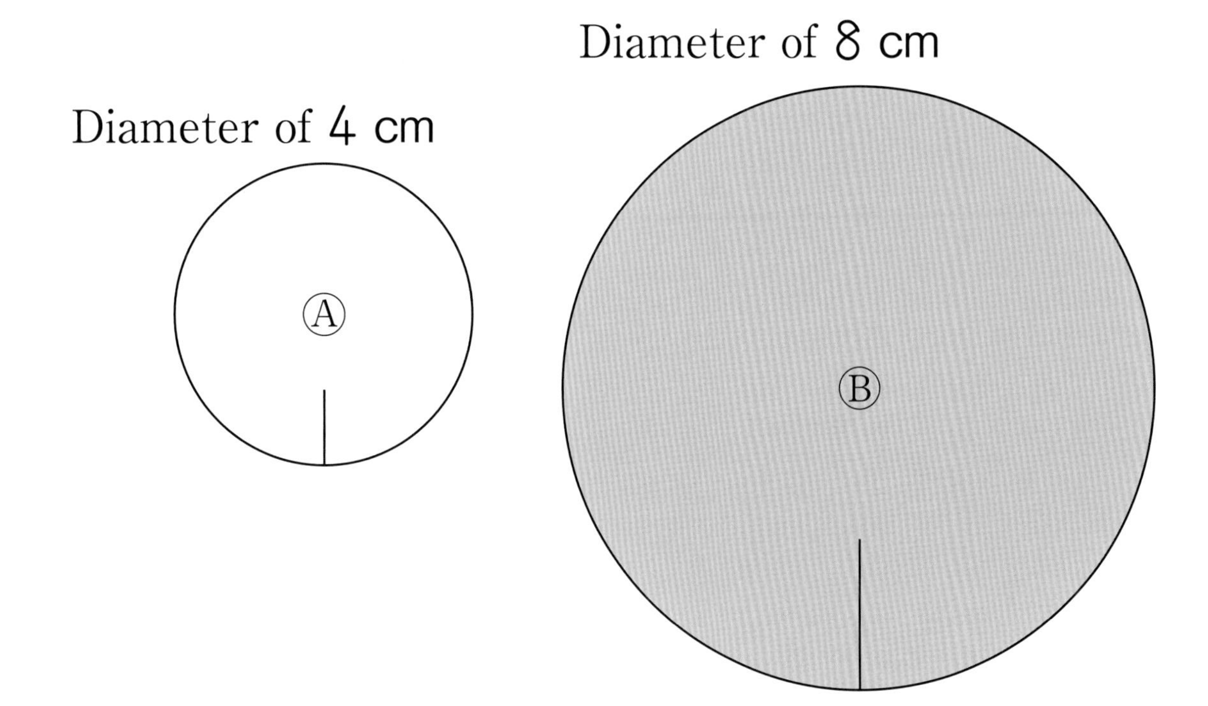
Diameter of 4 cm
Diameter of 8 cm
Ⓐ
Ⓑ

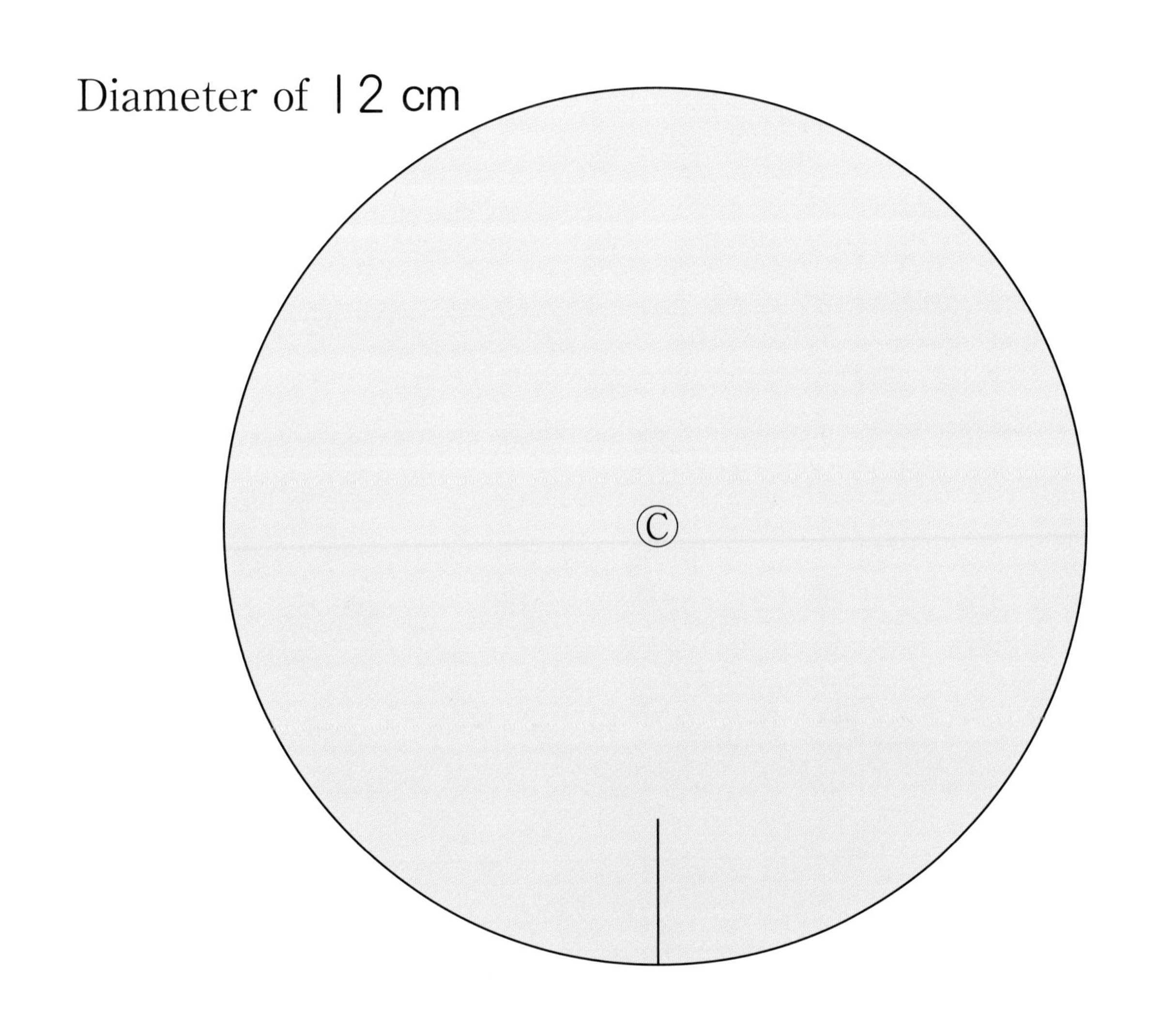
Diameter of 12 cm
Ⓒ

# Circumference Ruler

▼ will be used in page 74.

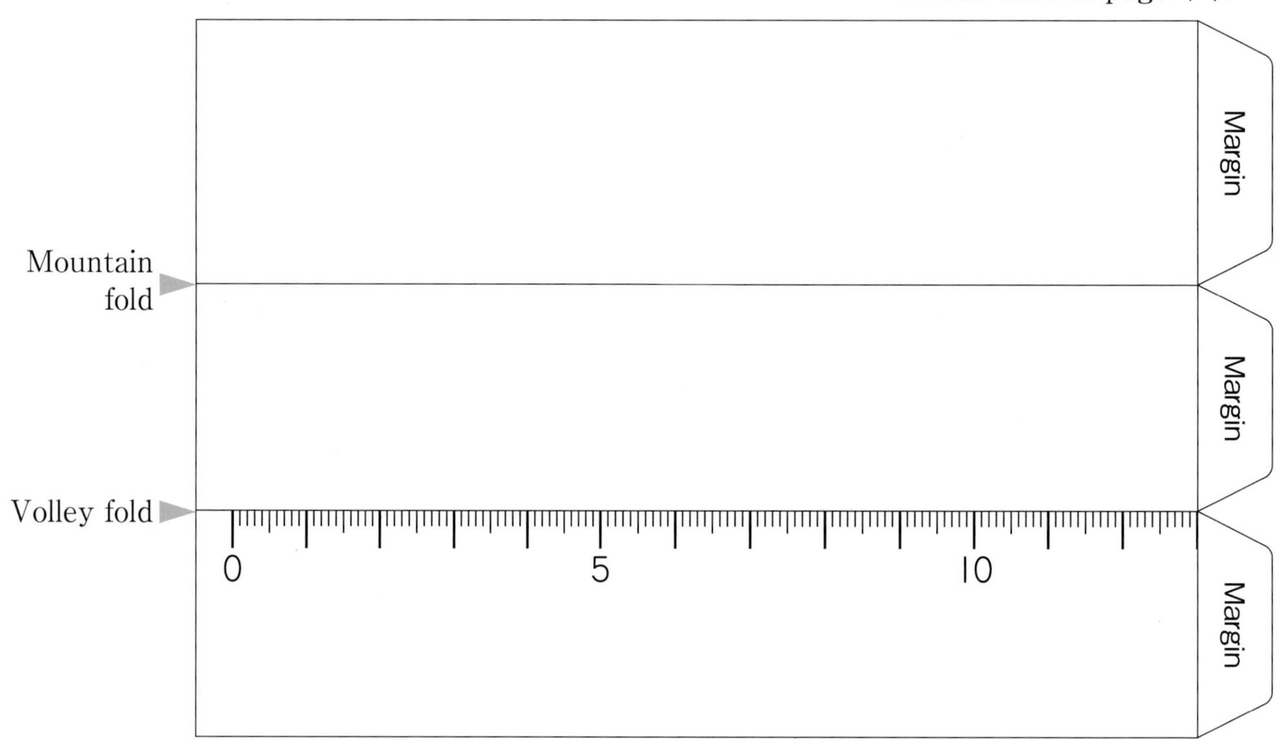


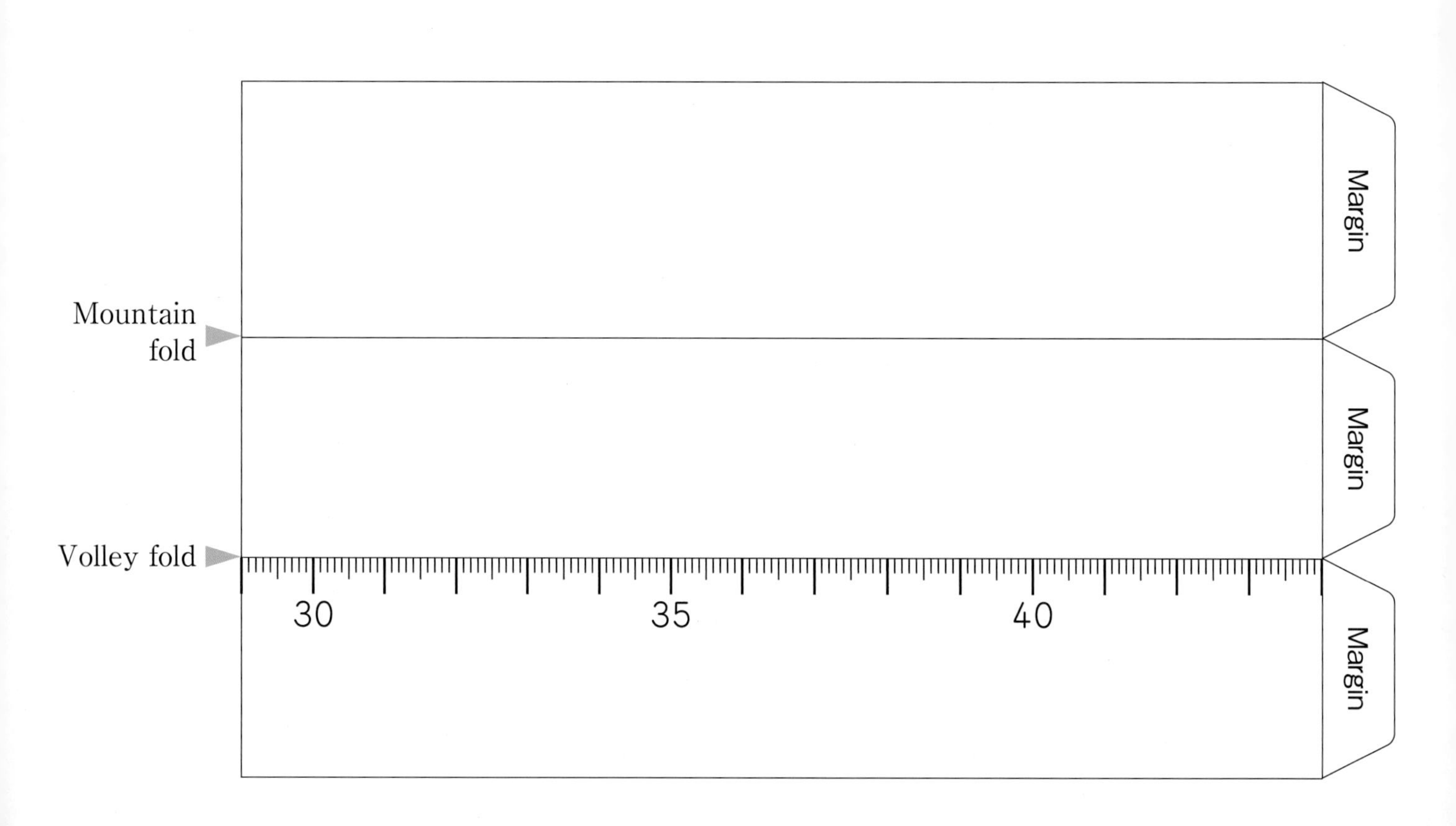

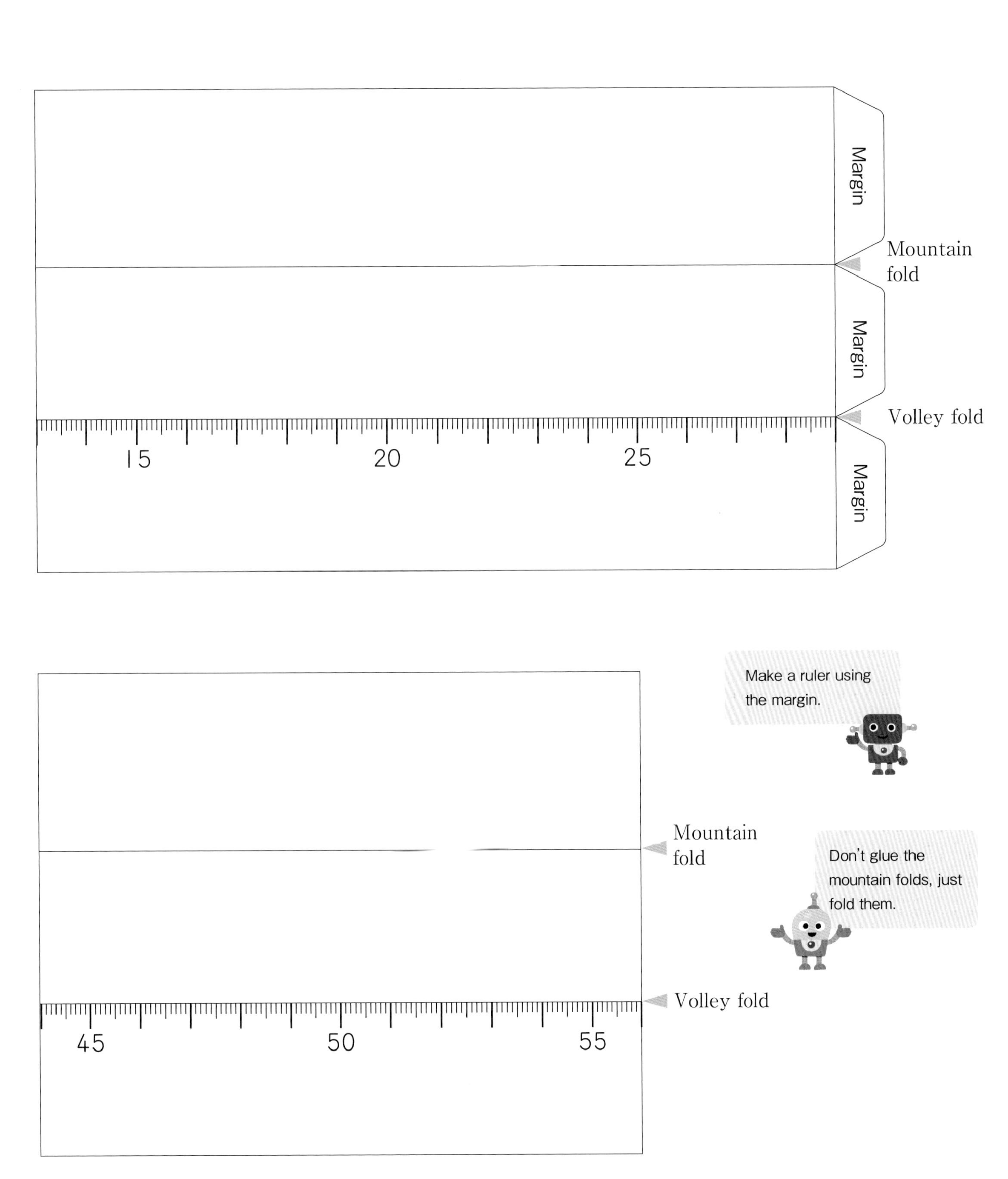
Margin
Mountain fold
Margin
Volley fold
15
20
25
Margin
Make a ruler using the margin.
Mountain fold
Don't glue the mountain folds, just fold them.
Volley fold
45
50
55

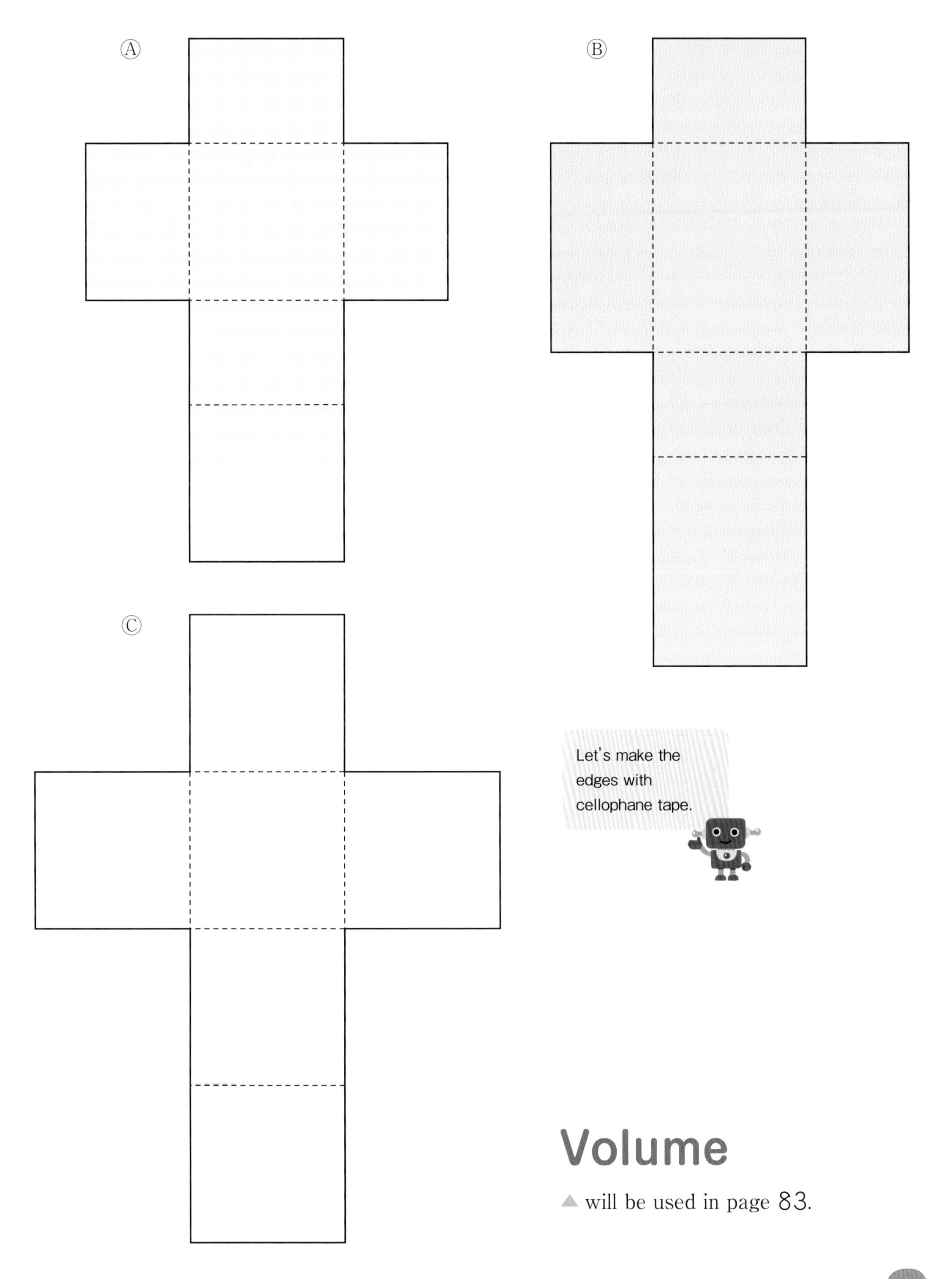


# Volume

▲ will be used in page 83.

# Memo

## Editorial for English Edition:

Study with Your Friends, Mathematics for Elementary School
5th Grade, Vol.2,  Gakko Tosho Co.,Ltd., Tokyo, Japan [2020]


---




### Translators:

Masami Isoda, Diego Solis Worsfold, Tsuyoshi Nomura (University of Tsukuba, Japan)
Hideo Watanabe (Musashino University, Japan)